DIRECT

D I R

HARPER
BUSINESS

An Imprint of HarperCollins*Publishers*

THE RISE OF THE

MIDDLEMAN ECONOMY

AND THE POWER OF GOING

TO THE SOURCE

E C T

KATHRYN JUDGE

HarperCollins books may be purchased for educational, business, or sales promotional use. For information, please email the Special Markets Department at SPsales@harpercollins.com.

FIRST EDITION

Library of Congress Cataloging-in-Publication Data
Names: Judge, Kathryn, author.
Title: Direct : the rise of the middleman economy and the power of going to the source / Kathryn Judge.
Identifiers: LCCN 2021054918 (print) | LCCN 2021054919 (ebook) | ISBN 9780063041974 (hardcover) | ISBN 9780063041981 (ebook)
Subjects: LCSH: Big business—United States. | Corporations—Political aspects—United States. | Capitalism—Moral and ethical aspects—United States. | United States—Economic conditions—1945-
Classification: LCC HD2785 .J83 2022 (print) | LCC HD2785 (ebook) | DDC 338.6/440973—dc23/eng/20220301
LC record available at https://lccn.loc.gov/2021054918
LC ebook record available at https://lccn.loc.gov/2021054919

22 23 24 25 26 LSC 10 9 8 7 6 5 4 3 2 1

To Judith Judge
1945–2020

| CONTENTS |

THE QUIET TRANSFORMATION

WALMART AND AMAZON are the two biggest companies in the country.[1] They generate more revenue than any of their peers, placing them first and second on the Fortune 500. They are also the biggest employers. At the end of 2020, Walmart employed 1.5 million Americans and another 1.3 million Americans worked at Amazon.[2]

More Americans shop at Walmart than at any other store or business. In 2016, 95 percent of Americans visited a Walmart. McDonald's came in second with just 89 percent.[3] In a 2021 survey of the most valuable brands in the world, Amazon came in first, beating out Apple, Google, Microsoft, Coca-Cola, and other household names.[4]

If the amount of money a company makes for its founders is any indication of its success, Amazon and Walmart could not be doing any better. Jeff Bezos is often the wealthiest person on the planet.[5] The Walton family—the heirs of Sam Walton, who founded Walmart—is the wealthiest family on the planet.[6]

Many of us have become so accustomed to relying on Amazon or Walmart that we take their size and omnipresence for granted. But there is much more to the rise of these giant middlemen, and the supply chains behind them, than meets the eye. They are just the most vivid examples of a much broader transformation in how the economy works, whom it works for, and where the real power now lies.

This book explains the rise of the "middleman economy"—characterized by powerful middlemen and long supply chains. It weaves together data, stories, and theory to show how the middleman economy took hold, the benefits it brings, and the dangers it poses. By comparing today's middleman economy with the very different ecosystem that grows out of direct exchange—between makers and consumers, investors and entrepreneurs—this book shows how much is at stake in the threshold issue of "through whom" we buy, invest, and even give. It demonstrates how increasing direct exchange and more modest shifts in that direction can help us lead richer lives and contribute to a more resilient, connected, and just economy. And it provides practical guidance about when to use middlemen, when to avoid them, and how to choose more selectively among them.

I have been studying middlemen and the growth of the middleman economy for more than a decade as a law professor at Columbia University. My field is financial regulation. My early work helps explain why the financial system circa 2007 was so fragile that underperforming subprime home loans could lead to widespread financial dysfunction and cause the entire economy to crater. My work also helps to explain why banks and other financial intermediaries so often succeed in pushing their regulatory agenda at the expense of ordinary Americans. Over time, I began to realize that the patterns I saw in finance arise in other domains as well, and for similar reasons.

Middlemen are the connectors. They help people and companies overcome the informational, logistical, and other hurdles that stand in the way of an exchange that would otherwise make them both better off. They are the retailers through whom goods travel when moving from producer to consumer. They are the banks through whom money moves from savers to people who want to buy homes and businesses that want to grow. They are the real estate agents who help people buy that home or convert their current home into money they can use for retirement. Many middlemen provide other valuable services as well, but so long as one of their core functions is to facilitate the movement of goods or money, they are middlemen.

Middlemen make the world as we know it possible. Thanks to middlemen, people living in the United States today can easily buy goods made on the other side of the world, build a diversified investment portfolio, order groceries from the comfort of their couch, search home listings while still sitting on that couch, and get the mortgage they need to buy their dream home. This helps to explain their rapid ascent to center stage.

But that is only part of the story. The very attributes that make middlemen good connectors also give them outsized power. In time, this enables them to expand their domains, entrench the need for their services, contort consumer decision making, and otherwise promote their interests at the expense of those they are meant to serve.

Perhaps the most defining characteristic of middlemen today is that in the process of connecting, they separate. They stand between consumers and investors, on the one hand, and the people and places behind the goods they consume and the projects they help to fund, on the other. Today, the very notion of "maker" has been displaced by supply chains that span continents. This often lowers costs but it also breeds new sources of fragility, undermines accountability, and leaves all of us more disconnected. This is the middleman economy in which we now reside. Learning to recognize the middleman economy for what it is can help us understand why we so often rely on giant middlemen, the ways middlemen affect what and how much we buy, the hidden impact of those decisions on other people and the environment, and the giant schism between the values many of us purport to hold and the state of the world we now inhabit.

Yet the inspiration to write this book came not from my research, but from a personal experience that allowed me to appreciate how things might be different.

———

Our second daughter was born with an unusual heart. So unusual, she had her first open-heart surgery at one week, and a second ten months later. I stirred with emotion—from joy and gratitude to isolation and powerlessness. We had incredible support from friends and family, but few understood what it was like to parent a child who might not see her first birthday.

In my umpteenth effort to learn more about her condition, I happened upon a salve more powerful than additional statistics: unexpected kinship. The source was a GoFundMe campaign for an infant in northern Canada with a similarly unusual heart. GoFundMe is a platform that allows people to ask friends, family, and strangers to support a dream or a challenge they are facing. The first family to whom I gave lived in a small town that was hours from the hospital where their daughter was being treated. As I read their story, I felt like I had found my people. They talked about pulmonary atresia, echocardiograms, and SpO_2 readings, the same once-foreign language that now surrounded me. Learning about their experience and making a small contribution made me feel slightly less alone, slightly less powerless.

Our daughter is now thriving, but I still visit GoFundMe to see updates from families I have helped and to give to new families in need. The comments, pictures, and email exchanges that often accompany donations suggest that I am not the only one to find comfort and community in these settings. This allowed me to appreciate anew how the opportunity to connect with the person on the other side of an exchange can make the exchange itself meaningful. At their best, direct exchange and platforms designed to encourage connection and communication can transform an exchange into an opportunity to forge new bonds, enjoy experiences we could never purchase in isolation, and see firsthand the impact of our actions on others. These dynamics stand in sharp contrast to the middleman economy, which is often designed to blind us to the people and places on the other side of an exchange.

Direct exchange and its kin are proliferating, for reasons both profound and practical. As the drawbacks of trying to navigate the middleman economy grow, so too do the benefits of finding a way around it. There has been an increase in true direct exchange, embodied in farmers' markets, creators selling homemade goods via their own websites, and comic cons replete with small-production graphic novelists hawking their wares. There has also been a proliferation of close cousins, including digital platforms such as GoFundMe, Etsy, and Kickstarter, and producers that bypass traditional middlemen, such as Warby Parker, Allbirds, and other direct-to-consumer companies. In providing consumers, investors, workers, entrepreneurs, and all of us as humans, an outside option, the rise of direct exchange can go a long way in shifting the balance of power even in a world still full of middlemen. Middlemen are not the enemy. The middleman economy is.

This book is my effort to meld the insights gained from years of research and from my own, often failed, attempts to forge a meaningful life to shed light on how we got here and where we might go next. It culminates in five principles that can help each of us regain control and contribute to a more sustainable economic system: intermediation matters; shorter is better; direct is best; know how middlemen make their money; and seek to build bridges. These principles can be used by individuals seeking more meaning in their lives, entrepreneurs looking for the next opportunity, and concerned citizens seeking to understand how we got here and how we can do better.

Part I compares the highly intermediated path most food follows from farm to table with a radically more direct alternative to bring to life each of the book's major themes. Part II examines the rise of the middleman economy in other domains, showing how middlemen both enable and are strengthened by the rise of long, complex supply chains, and how both developments—the core of the middleman economy—have changed the nature of work and the structure of society. Part III exposes the dark side of the middleman economy. It reveals that the very infrastructure, expertise, and relationships that make middlemen so useful also enable them to further their own interests at the expense of the rest of us. It further reveals how the complex supply chains that create short-term efficiencies also breed new sources of fragility, and tend to break down at the worst possible times. Part IV examines how direct exchange can serve as the cornerstone for a better system. By explaining why middlemen are so useful, the hidden dangers that they pose, and the distinct benefits of going to the source, the four parts work together to show the trade-offs at stake in the current system, the very different benefits and costs of simpler, more direct modes of exchange, and where change can have the greatest impact.

The last chapter integrates the core insights from the rise of the middleman economy and the transformative power of direct exchange into the five principles. Middlemen are so omnipresent in our lives that it is easy to forget that we have a choice. Adding more direct exchange to our individual economic diets can provide the sense of connection and appreciation of our inherent interconnectedness that get lost amidst the many separations that middlemen perpetuate. Government interventions can provide similar benefits for society, counteracting excessive concentrations of power while promoting more resilience and accountability.

The analysis here provides fresh support for efforts to revive competition policy, particularly when the actors involved are middlemen. It also illuminates how much more the government could do to support bottom-up change. Small makers will never posess the fleets of trucks, mountains of customer data, and other advantages enjoyed by Amazon, Walmart, and other giant middlemen. From ensuring the viability of the U.S. Postal Service to setting aside public spaces for craft fairs, the government can play a critical role helping to create and subsidize the infrastructure needed to make direct exchange viable. Just like top-down efforts to keep monopolies in check, these policies help level the playing field and make it easier for people to opt out of the dominant system.

A note to fellow academics and the insatiably curious: We are in the midst of an overdue reckoning regarding the virtues and limits of using a "neoliberal" frame for understanding the institutions, public and private, that can best promote individual and collective flourishing. Some continue to use efficiency-based reasoning to defend market-based mechanisms against state interference, others are using the elasticity of such reasoning to show why markets so often produce bad outcomes and why more government intervention is needed, and yet others—long present but growing in influence—suggest that the entire paradigm is so flawed that it ought to be dismissed in its entirety. I take the unusual approach for an academic, at this historical moment, of integrating insights from economics with an understanding of the need to move beyond such framing to make real progress on the challenges we now face.

Insights drawn from economics are key to explaining both the rise of middlemen—how intermediation can enhance productivity and allocation—and why the same changes that increase efficiency in the short run set the stage for a host of "market failures" down the line. Even if the only aim is to maximize output relative to scarce inputs, the contours of today's middleman economy are far from optimal.

Yet an excessive focus on matters such as efficiency is key to explaining how we got here, and why lawmakers allowed such concentrations of power and fragile schemes to take hold. Moving past a scarcity mindset also reveals how much more is at stake, from the quality of our interactions to the nature of work and our sense of community. Weaving together insights from competing paradigms produces a richer—even if less theoretically pure—account of why intermediation matters, why the world looks as it does, and how we can all help forge a better tomorrow. And if "no problem can be solved from the same level of consciousness that created it," this may be the only way to open up new and better horizons.[7]

PART I

TRACING FOOD FROM FARM TO TABLE

CHAPTER 1

THE HIDDEN COST
OF CONVENIENCE

N 2011, CANTALOUPE contaminated with *Listeria monocytogenes* killed more than thirty people and sickened 140.[1] All of the cantaloupe came from a single farm in Colorado, yet the victims were spread across twenty-eight different states. Supply chains that carry food far from its origins can also carry pathogenic bacteria. And the more food travels, the harder it can be to trace those origins, enabling the danger to spread even further.

These dynamics were on full display that same year in Europe. An *E. coli* outbreak started in Germany in early May 2011, but no one could figure out the source.[2] German health officials warned citizens not to eat raw tomatoes, cucumbers, or lettuce based on early indications that fresh produce might be to blame.[3] Yet new cases continued to emerge. Hundreds, and then a few thousand people, fell ill. Many Germans experienced a pervasive sense of anxiety.[4] "I am treated like a potential murderer," one retailer said, "simply because I sell cucumbers and tomatoes."[5]

On May 26, German health officials identified cucumbers from

Spain as the likely culprit.[6] Consumers across Europe started avoiding produce grown in Spain. Spanish farmers saw their fresh produce wilt, unsold, leading to losses of at least 50 million euros and potentially four times that amount ($72 to $280 million, at that time). Farmers in Belgium, Bulgaria, France, Portugal, Switzerland, the Netherlands, and Germany also suffered, as many Europeans remained fearful of any fresh produce.[7] Tourists who had planned to visit and sports teams scheduled to compete in Germany canceled their trips to avoid the risk of contaminated food, increasing the economic fallout.[8]

Eventually, public health officials realized that neither cucumbers nor Spanish farms had played the slightest role in the outbreak. The real source of the *E. coli* was far closer to home: salad sprouts grown right in Germany.[9] Unfortunately, by the time the true origin was suspected, most of the tainted sprouts had been consumed to the detriment of those who ate them. Ultimately, the outbreak caused 4,000 people to fall ill and resulted in 54 deaths.[10] The full economic damage, much less the fear and anxiety the episode triggered, has never been quantified.

Although such outbreaks remain rare, they illustrate the dangers of producing food in large volumes and then distributing it across a continent and beyond. Even though the sprouts that posed such a danger were grown right in Germany, the fact that those who got ill were simultaneously consuming produce and other food from so many different places made it difficult for public health officials to identify the true source of the problem. More people got sick, more people died, and more farmers suffered as a result.

There are also signs that the same sprout seed stock—originally from Egypt, imported to Europe via Rotterdam, with some then going to Germany and some reaching France via the United Kingdom—caused a smaller-scale *E. coli* outbreak in France later in the spring of 2011. But again, the complexity of the chains through which the seeds and other produce traveled and the ways that complexity impeded the public health investigation precluded researchers from being able to do the type of testing needed to confirm this suspicion.[11]

Even for those who avert death, the body's response to food

poisoning can cause lasting damage. This is a lesson that Stephanie Smith of Minnesota learned the hard way.[12] Stephanie was twenty-two when she started experiencing stomach cramps. She managed to make it through her workday, teaching children to dance, before landing in the emergency room. Once there, her speech became slurred, her kidneys failed, and she started suffering from seizures. At the advice of doctors, her mother consented to Stephanie being drugged into a coma and flown to a Mayo Clinic hospital. Stephanie survived, but her cognitive abilities remain permanently impaired. She is unlikely to ever dance or walk again.

Doctors eventually identified *E. coli* in a hamburger she had eaten at a family barbecue as the source of her troubles. The meat was traced to Sam's Club, which had purchased the hamburgers from Cargill. But Cargill was not the producer of the contaminated meat. It was a middleman. Cargill is a food processing behemoth. Among the ways it makes money is by acquiring, processing, and packaging beef from various slaughterhouses, which it then sells to retail middlemen such as Sam's Club. Because the meat is mixed together, even Cargill was not able to readily identify the slaughterhouse that had actually been the source of the tainted meat.

Stephanie's family sued Cargill. After a couple of years of litigation, the company agreed to settle the case for an undisclosed amount. Cargill never publicly accepted that it was at fault for Stephanie's illness. Settling without a trial kept the decisions that had allowed the contaminated burger to reach Stephanie hidden from public view.[13] Exceptional investigative reporting by Michael Moss is the only reason Stephanie's case got any public attention. And even Moss, a Pulitzer Prize–winning journalist, struggled to overcome government resistance and other challenges in his quest for answers. He eventually learned that the burger that paralyzed Stephanie "had been an amalgam of various grades of meat from different parts of the cow and from multiple slaughterhouses as far away as Uruguay."[14] In a world of middlemen, some answers remain elusive. Accountability, both legal and moral, suffers.

Middlemen and long intermediation chains are not directly responsible for most foodborne illness. Cargill didn't want Stephanie to be ill. None of the middlemen that helped distribute the tainted sprouts in Germany bore any ill will toward those who bought their products. In fact, given the legal risks, most middlemen want the products that they help distribute to be free from harmful contaminants.

Yet middlemen need not be evil, or even indifferent to harm, to be part of the problem. Although there are a growing number of efforts to improve the traceability of food from origin to consumption, these efforts are costly and remain far from perfect. Recent research suggests most models of how to implement tracing systems still fail to capture the messy realities of how food moves through the economy.[15] The aggregating, mixing, and subsequent splitting that often happens at multiple nodes along the way make actual tracing almost impossible much of the time. This is why even a large, sophisticated middleman like Cargill had such a hard time tracing the origins of the meat that moved through its own processing plant.

Long and complex supply chains are one of the defining features of the middleman economy. The way meat from cows slaughtered across the United States and Uruguay ended up potentially mixed together in hamburgers for sale at a Sam's Club in Minnesota exemplifies how today's supply chains work. They often accrete slowly and seem to create efficiencies as they grow and evolve. But, too often, the complexity also leads to fragility, unknowns, and a dearth of accountability. *E. coli* outbreaks that spread further and last longer because of traceability problems are just one example of the bad things that can result.

Another common challenge is that people end up buying food produced in ways that they would find unethical—*if* they actually knew what was going on. Yet because large middlemen and complex supply chains blind consumers to unpleasant realities, they continue to support practices inconsistent with their values. One example causes me particular pain.

THAT CHOCOLATE BAR

I love chocolate. It takes self-control for me to resist stealing cheap treats from my kids' Halloween baskets, and I don't always succeed. For a long time, I was at peace with this guilty pleasure, as I lived under the sheltered delusion that the negligible effect on my health was the primary human impact at stake.

I have since learned otherwise. The two countries that grow the most cocoa, Cote d'Ivoire and Ghana, are infamous for allowing child and forced labor, poor working conditions, and extremely low wages. In 2001, growing concerns about these conditions motivated the U.S. House of Representatives to pass a bill that would have given the Food and Drug Administration $250,000 to develop "slave-free" labeling requirements for chocolate products.[16] The legislation was abandoned only because Hershey, Nestlé, Mars, and the other chocolate companies fought back. With the help of two trade associations looking out for their collective interests, they hired a team of lobbyists, including former senator Bob Dole, and succeeded in defeating the bill.

At first, these companies may appear to be makers, not middlemen. And it is true that in mixing cocoa from West Africa with sugar, milk, and other ingredients, shaping those confections into bars, and putting them in pretty wrappers, they are very much producing something new. Yet much of the value they provide comes from their ability to procure ingredients from around the world and then deliver finished products to retailers and consumers. These companies thus embody an important and pervasive feature of today's middleman economy: actors that are both makers and middlemen. Without belittling the other functions that they play, recognizing chocolate giants as part middlemen is key to mapping the increasing number of nodes involved in production and the reasons we are often so disconnected from and ignorant of the people and places behind the goods we consume.

One way giant chocolate middlemen helped secure their victory on Capitol Hill was by promising they would do better. In 2001, along with members of Congress and nonprofits devoted to eradicating child labor and improving working conditions, these companies entered into an agreement that was supposed to usher in a new era. The agreement included numerous and detailed provisions for reducing and eventually eradicating troubling labor practices in chocolate supply chains. The heads of Nestlé, Hershey, and Mars all signed the document, signaling their purported commitment to meeting these goals.[17]

Two decades later, child labor remains rampant. The initial draft of a 2020 report funded by the U.S. Department of Labor and conducted by researchers at the University of Chicago found widespread problems in Cote d'Ivoire and Ghana, which continue to produce roughly 60 percent of the world's cocoa. According to the draft report, the proportion of children (ages 5 to 17) in the area involved in the production of cocoa actually increased from 31 percent in 2008–09 to 41 percent in 2018–19.[18] The last published study, which the Department of Labor conducted in conjunction with researchers from Tulane in 2015, also found an upward trajectory. According to that report, in Cote d'Ivoire and Ghana alone, 2.3 million children were working in cocoa production, and more than 2 million of those child workers faced hazardous working conditions.[19] Sexism also appears to be widespread, with woman providing more of the labor while receiving far less compensation for their efforts.[20]

Extensive on-the-ground research conducted by the Corporate Accountability Lab sheds further light on the problems endemic to cocoa production today. A report summarizing the Lab's research and findings suggests most cocoa farmers are paid far too little for the goods they produce. This results in rampant poverty in addition to accentuating the pressure on farmers to hire the cheapest workers possible, such as children and those forced into labor. Most chocolate middlemen also remain unable to identify the origins of much of the cocoa they use. According to the *Washington Post*, in 2019, Mars could trace the origins of only 24 percent of its cocoa. Hershey and Nestlé did somewhat better, but both could trace less than half of their global cocoa supply to the farms where that cocoa was grown.[21] Godiva is even more derelict. A cohort of nonprofits gave Godiva the "Rotten Egg Award" before Easter 2020, "for failing to take responsibility for the conditions in which its chocolates are made, despite making huge profits" and providing almost no information about the origins of the cocoa in its products.[22]

According to the Corporate Accountability Lab, one reason the past two decades yielded so few improvements is the proliferation of third-party certification schemes.[23] Rainforest Alliance, Fairtrade International, Fair Trade Certified, and Cocoa Life are among the third parties that purport to "certify" cocoa.[24] There is some evidence suggesting that when cocoa is certified, farmers do earn a slightly higher income from selling it. Nonetheless, when Corporate Accountability Lab researchers engaged in extensive consultations with the farmers, they found "virtually imperceptible differences between certified and uncertified farms in terms of living incomes, poverty, education, access to healthcare, farmer bargaining power, or access to information."[25]

Buying "certified" cocoa allows large chocolate companies to signal virtue, creating a semblance of concern for the well-being of those who labor to produce the cocoa on which they depend. But as reflected in the persistence of problems on the ground and the widespread ignorance among chocolate companies regarding the true origins of the cocoa they use, efforts to tack on labels without shortening supply chains rarely suffice to bring about meaningful accountability.

Concerned consumers and nonprofits have sought other avenues for promoting transparency and accountability. In a series of lawsuits, consumers sued Nestlé and other major chocolate middlemen. The consumers claimed they never would have bought the chocolate had they known more about the conditions of the laborers farming the cocoa that chocolate contained, so, under California law, the companies had an obligation to disclose this information to prospective buyers. As someone who loves chocolates and abhors child labor, this seemed like a reasonable claim to me.

The court disagreed. The federal court that issued the final ruling in the suit acknowledged that "child labor and slave labor are modern-day scourges" and that chocolate companies such as Nestlé may well "benefit from that illicit labor."[26] Nonetheless, the court ruled that Nestlé and other giant chocolate middlemen have no affirmative obligation under the California law to "label their goods as possibly being produced by child or slave labor." The court's view was that the central function of a chocolate bar lay in its physical characteristics—how good it tastes—not the supply chain behind it. Thus, no matter how reprehensible the labor practices may have been, they can remain hidden from consumers, even in a consumer-friendly state like California.[27]

As a result, each time a shopper picks up a candy bar at the checkout aisle, he is indirectly supporting a system that exploits women and children while enriching the giant middlemen that have worked to entrench that system. The retail middlemen that put together lavish Halloween and Easter displays are also careful to hide these realities from view, and are helped by the long distances and many layers separating African cocoa farms and Americans stocking up on sweet treats.

Third-party certification schemes, from organic labels on food to sustainability labels on clothes and ESG (Environmental, Social, and Governance) labels on investments, are proliferating. Their growth reflects just how much today's consumers and investors care about issues beyond low prices and high investment returns. As chocolate's travails illustrate, however, these schemes can obscure as much as they illuminate. On their own, they are no match against today's long supply chains and the middlemen that profit from hiding unpleasant realities from view.

HOW FOOD IS GROWN

Given the many drawbacks of having food travel so far and through so many layers between farm and table, it may seem odd that this has become the norm. The primary benefit of this highly intermediated system is that it saves consumers money. In 1900, the average U.S. household spent 43 percent of its income on food. Food was the single biggest expenditure, exceeding housing, clothing, and health care.[28] By 1950, food was just 30 percent of the budget; and by 2003, that figure was down to 13 percent of the moneys spent by the average U.S. household.[29]

As chocolate reveals, some of these cost savings arise from imposing inhumane working conditions on faraway farmers and then shielding consumers from those realities. Yet much of the cost savings have come from changes in how people farm and efforts to harness the benefits that can come from the scale and hyper-specialization that middlemen enable.

These dynamics are on full display in Illinois, home of 75,000 farms that collectively cover three-quarters of the state's land. Laura operates one of those farms, on the outskirts of the town of Pontiac. Like many in the area, Laura grew up farming. Her father, Frank, was a farmer, as was his father before him. Farming is all that Laura ever really wanted to do. Most of the land that Laura farms is land that she or a family member owns.

Laura's farm is family controlled, but it is a high-tech, large-scale operation. She does not plant an array of crops. She plants two: corn and soybeans. Rather than selling her crops for immediate consumption, she dries both in large silos and then sells when she thinks the price is right or she needs the cash. Until recent trade tensions, she saw China as her biggest market.

I know this because Laura is my mom's first cousin. We grew up visiting her farm. I will never forget the first time I visited Laura as a child. It was during harvest, a task she accomplished with the aid of a combine harvester. I was awed by the massive size of the machine, with tires that dwarfed my six-year-old self. The way it appeared to methodically plow down row upon row of corn was like nothing I had seen before. But what *really* impressed me was the cockpit. After climbing up a ladder so tall that I was scared to look down, I peeked into a deluxe compartment with a cushy black seat, air-conditioning, and a state-of-the-art sound system. I was convinced that farming was a high-end, and rather cushy, occupation.

Laura still uses large, complex machines like combines with increasingly advanced technology for all aspects of planting and harvesting. On our last visit, my older daughter was just as awed as I had once been when we watched the combine in action. This machinery is just one of the ways that Laura is deeply enmeshed in what many call the "agricultural industrial complex." The way in which Laura farms, and how different it is from how her grandfather farmed the same land, brings to life the way middlemen do much more than connect. As intermediation evolves, so does the competitive landscape, and that both enables and necessitates fundamental changes in how food is grown and goods are made.

The ways cousin Laura approaches farming is representative of most U.S. farmers today. In 1935, there were approximately 6.8 million farms in the United States. That number has shrunk to 2 million even as the population has swelled.[30] Productivity and revenue are up, but so are costs.[31] State-of-the-art combines do not come cheap. Laura's choice of crops is also representative: Soy and corn are the two most planted crops, and collectively account for 43 percent of the nearly $200 billion worth of U.S. crops, including fruit, vegetables, nuts, cotton, and hay, grown and sold in the average year.[32]

Technology is pervasive, and not limited to the machines used for planting and harvesting. A 2018 study found that a full third of U.S. farmers have used drones, either on their own or through a third-party provider, for activities such as monitoring crops and deploying fertilizer.[33] A report from PricewaterhouseCoopers suggests that agriculture-related use of drones will eventually be a $30 billion market, making agriculture the second-largest commercial market for drones.[34]

The trend away from producing for local consumption is also the norm. Agricultural imports and exports have been rising steadily, more than doubling in value between 2000 and 2015.[35] Much of this trade occurs within North America, but there is also a high volume of overseas imports and exports. According to the U.S. Department of Agriculture (USDA), "East Asia—led by China, Japan, and South Korea—was the largest market," collectively buying 34 percent of the grain and other agricultural products the United States exported in 2018.[36] About 50 percent of the soy and more than 20 percent of the corn grown in the United States will be exported.[37] The United States also imports a lot of food. The Food and Drug Administration estimates that 80 percent of the seafood, 50 percent of the fresh fruit, and 20 percent of the fresh vegetables Americans eat come from abroad.[38] Even when consumed in the United States, food travels an average of 1,500 to 2,500 miles from the site of production to consumption.[39]

Middlemen have facilitated each of these developments: In a world where food can travel more easily and cheaply, more of it travels and it travels farther. This is why cantaloupe from Colorado ended up sickening people in twenty-eight different states. This is why people have to make an effort to "eat local," despite having no other choice for most of human history.

Thanks to middlemen, Laura doesn't have to figure out who most wants her crops and then find a way to connect with the livestock producers in China who feed it to their animals. Nor does she have to figure out how to transport her grain from Illinois to the Far East. All of this is undertaken instead by a complex web of middlemen. There are advantages to this system. Laura can devote all of her attention and resources to farming and invest in the vast amounts of land and fancy equipment she needs to farm as she does. She is able to devote so much money and effort into becoming very good at one narrow thing because middlemen do the rest for her.

But hyper-specialization is not a choice for Laura. It is a survival technique. She competes with producers the world over. Corn and soybeans are commodities. If Laura incurs extra costs to grow her crops in a way that is more respectful of the land or on a more modest scale, she cannot readily pass those costs on to her customers. She must take whatever price the market sets and find a way to make it work. When lacking the cash for necessary land or equipment, she takes on debt. And again, she is not alone. According to the USDA, farm real estate debt is expected to reach $287 billion in 2021, and other debt owed by farmers is likely to exceed $150 billion.[40] This helps to explain why farm bankruptcies in the United States have been increasing, in general, since 1980.[41] More than 300 farms have been forced into bankruptcy each year since 2013. In recent years, the figure has exceeded 500 farms annually. Despite what I thought as a child, farming is not a road to riches or even economic security.

Moreover, the costs of the way most farms operate today are not just borne by the farmers. American farms today use forty times as much nitrogen-based fertilizer as they did seventy-five years ago, when my great-grandfather was tending to the fields.[42] A 2019 study published in *Nature* found that the adverse impact of the production of corn (maize) on air quality is associated with 4,300 premature deaths in the United States each year.[43] Much of the pollution and loss of life is concentrated in just five "Corn Belt" states, including Illinois. Modern farming has helped increase production—enabling American farms to remain competitive in an increasingly global marketplace and providing cheap food for consumers—but other people and the planet pay much of the price.

Nor are these issues unique to the United States. Seventy percent of the fresh water taken from the earth is currently used to support agriculture.[44] More than a quarter of greenhouse gas emissions arise from agriculture and food processing.[45] Twenty-four billion tons of fertile soil are lost each year due to practices associated with industrial agriculture, detrimentally degrading a full third of the world's arable land.[46] None of this can be sustained.

WHERE THE REAL POWER LIES

Cousin Laura's farm may be large and industrial by historical standards but it is still a family farm. So are the great majority of farms in the United States. As of 2017, just 2.2 percent of all farms in the United States were not family owned.[47] Most of these farms are small or midsized affairs, with the majority of American farmers earning less than $150,000 from farming each year.[48]

Shift the focus from farmers to processors, like the chocolate companies we saw earlier, and the picture looks very different. The top ten food and beverage companies earned $450 billion in revenue in 2019, more than the next thirty companies combined.[49] These companies, such as Nestlé, Mars, PepsiCo, and the Coca-Cola Company, represent the second defining feature of today's middleman economy: very large, very powerful middlemen. That they are makers in addition to being middlemen does not negate the role they play helping to connect and separate those who grow food from those who consume it.

Although these may seem like household names, their full scale and scope is often hidden from view. I was shocked when I learned that Blue Bottle Coffee is not an independent company but is instead controlled by the largest food middleman: Nestlé. Odds are, even if you have never heard of Blue Bottle, you too have consumed a Nestlé product without realizing it. If you have ever picked up a KitKat, Gerber baby food, Cheerios, Stouffer's, Hot Pockets, Häagen-Dazs, Carnation, or some Purina or Friskies for your pet, you were buying a Nestlé product. At one point, Nestlé owned Poland Spring, San Pellegrino, and Perrier. PepsiCo too makes far more than soda. It also owns Quaker Oats, Gatorade, Lipton, Lay's, Doritos, SodaStream, and a total of twenty-three different brands that each produced over $1 billion in revenue in 2019.

That food middlemen actively promote an appearance of diversity that obscures just how much control lies in the hands of just a handful of giants comes through in a visit to Blue Bottle's website. Despite the fact that the company is controlled by Nestlé—which means Nestlé could replace the entire board of directors and management team at any time—there is no reference to Nestlé anywhere on Blue Bottle's home page. Even someone who clicks on "our story" to learn more about the company finds only a heartwarming tale about the company's quaint beginnings in Oakland, California.[50] Nestlé's dominance, and the fact that Nestlé is the primary beneficiary each time someone shells out four dollars for a cup of Blue Bottle coffee, is conspicuously hidden from view.

The food world is full of similar efforts by food middlemen to feed people saccharine substitutes for the nourishing, authentic sweetness they crave. This is why there are pictures of farmers and wheat fields grazing so many cereal and cracker boxes, even though farmers receive only a fraction of what consumers now pay for food. This is why so many food products use typeface that mimics handwriting even though it is clearly printed by a machine. Studies have found that using a handwritten as opposed to a typewritten font "increases perceptions of human presence," making people view a product more favorably and creating more of an "emotional attachment to the product."[51] Just as with Blue Bottle, however, the pictures of farmers and typeface that evokes a sense of humanity are merely superficial glosses. They are techniques used by massive middlemen that can afford to investigate and are not hesitant to exploit the behavioral biases of the customers they are supposed to serve.

Cargill, the key middleman behind the burger that debilitated Stephanie Smith, embodies just how large and powerful these middlemen have become. Each year, Cargill transports 70 million metric tons of food internationally.[52] At any given moment, it has an average of 650 vessels crossing the world's oceans. As the company explains in its 2018 annual report: "Our global presence in agricultural supply chains, deep understanding of markets and vantage points across sectors empower us to provide unique offerings to our customers."[53] This is true. What Cargill doesn't say, but surely knows, is that its "global presence," "deep understanding," and "vantage points" across so many sectors also allow it to shape policy and evolving market structures in ways that serve its interests at the expense of customers and the public at large.

IMPACT ON POLICY MAKING AND MAKERS

In theory, government is supposed to look out for the interests of the public. In practice, policy making often veers far from this ideal, and today's powerful middlemen are adept at using this gap to their advantage.

One well-known challenge with policy making in practice is that elections have become costly affairs and elected officials rely heavily on campaign contributions in order to get reelected. This tilts the playing field, and access to elected officials, toward those with money to give.[54] The big food middlemen know this and give generously. According to OpenSecrets.org, which gathers data to enable greater public accountability regarding lawmaking, agribusinesses including food middlemen increased giving to $118 million in campaign contributions during the 2016 election cycle—more than the industry had ever given previously.[55] It then broke that record in 2020, when contributions totaled more than $186 million, three times the amount given just twenty years earlier.

Even when lawmakers are well intentioned and without conflict, middlemen still have a remarkable ability to distort policy in self-serving ways. This is where the many informational and positional advantages that Cargill bragged about in its annual report come into play. Elected officials have to address way too many different issues to become a true expert in any of them. And even policy makers at specialized agencies typically know far less than large middlemen about how a specific market works. Middlemen know this, and take advantage accordingly. They use their wealth of knowledge to highlight the advantages of policies from which they benefit and the drawbacks of policies they dislike. What makes middlemen particularly pernicious is their ability to use their understanding of consumers, farmers, and other key groups to spin plausible tales of why a policy they don't like would also adversely affect the very people it is meant to help. Cargill, for example, has expended significant resources setting up an "advocacy website" that "show[s] how farmers, workers and consumers are better off thanks to global trade."[56] The name of the site—fedbytrade.com—sums up the message, which is then brought home through well-designed graphics and moving images of farmers and people enjoying good food that, by implication, they can access only because of cross-border trade. As is often the case, Cargill is providing information that is truthful but incomplete, focusing on the benefits and eliding drawbacks. Although far from the only factor, these types of efforts help to explain the rapid spread of policies enabling freer trade despite the lack of corresponding reforms to help those disadvantaged by such policies.

Exacerbating the challenge, the food industry regularly sponsors conferences, funds academic research, and otherwise shapes what academics study and the findings they are inclined to present publicly.[57]

Similar challenges arise in other jurisdictions. According to a 2014 study by the Corporate Europe Observatory—a nonprofit dedicated to exposing the influence of corporate lobbying on European Union (EU) policy making—agribusinesses, including middlemen like Nestlé, PepsiCo, and Cargill, were the most aggressive of all of the industries engaged in lobbying to reduce barriers to cross-border trade. The group found that lobbyists from agribusiness companies had more contacts with the trade department of the European Commission than lobbyists from the pharmaceutical, chemical, financial, and auto industries put together.[58]

The influence of food middlemen in shaping research and policy making came to the fore in a public health initiative focused on animal husbandry—the branch of agriculture focused on raising animals for meat and other products. In the mid-2000s, the Pew Charitable Trusts and the Johns Hopkins Bloomberg School of Public Health formed a commission of fourteen former policy makers, animal health experts, public health experts, and other professionals to study animal agriculture in the United States.[59] After two years soliciting technical reports, visiting diverse animal farms, and engaging in extensive learning about how animals destined for consumption are born and bred, the commission issued a report that made it clear that meat middlemen had become far too powerful, with devastating consequences for farmers, consumers, and animals.

On the positive side, the Pew Commission acknowledged that changes in how animals are raised, sold, and processed over the last fifty years have produced meaningful cost savings. In 1970, Americans spent 4.2 percent of their income on meat and poultry. By 2005, they were spending half that much, despite consuming even more meat.[60]

Yet, these savings came at a cost. The commission found that the farmers who raise the animals often do not own those animals. Instead, they are owned by the middleman, which also "controls all phases of production, including what and when the animals are fed."[61] When coupled with the concentration among integrators—the role played by the middleman—the net result is that there are "a small number of companies overseeing most of the chicken meat and egg production in the United States."[62]

The 2008 report further showed that as middlemen increased their control, the harm inflicted on other people and the environment increased. The problems included an "increase in the pool of antibiotic-resistant bacteria because of the overuse of antibiotics; air quality problems; the contamination of rivers, streams, and coastal waters with concentrated animal waste; animal welfare problems, . . . and significant shifts in the social structure and economy of many farming regions throughout the country."[63]

Yet the commission did not just learn about how bad things had gotten, they also saw firsthand the forces that contributed to these problems and the failure of regulators to address them. They laid the blame squarely at the feet of an "an alliance of agriculture commodity groups, scientists at academic institutions who are paid by the industry, and their friends on Capitol Hill."[64] The commission even found its own work stymied, as authors from whom the commission sought information had their funding threatened. Overall, the commission "found significant influence by the industry at every turn:

in academic research, agriculture policy development, government regulation, and enforcement."[65]

The policy response to the study, or rather lack thereof, is further evidence of how effectively food middlemen use their many vectors of influence to shape the rules of the game. Despite the study's disturbing findings and the positive press it garnered, a follow-up study five years later found minimal progress addressing the issues the report exposed.[66] According to the subsequent report, the Obama administration failed to engage with most of the recommendations, regulators "acted regressively in their decision-making and policy-setting procedures," and Congress sought to impede meaningful reform.[67] The USDA also ignored a recommendation from its scientific advisory committee to include recommendations to limit the consumption of red and processed meat.[68]

CONCENTRATION

The situation remains dire. When law professor and activist Zephyr Teachout set out to write a book about companies she saw as so big and powerful that they threaten public welfare, she started with the chicken industry. Consistent with the findings of the Pew Commission, Teachout found that just three companies—Tyson, Perdue, and Pilgrim's Pride—had bought up so many of their smaller competitors that they collectively "are responsible for buying and selling almost every chicken in America."[69] Concentration is nearly as high among other meat middlemen. For example, just four processors control 80 percent of the beef market.[70]

"I get phone calls continually about suicide," a third-generation Alabama chicken farmer told Teachout. Two of his fellow chicken farmers had already taken their lives, and his own health was ailing. As Teachout explains, the high degree of concentration among the middlemen that control the chicken market results in a bargain that puts all of the control in the hands of the middlemen while forcing farmers to bear much of the risk. Being a farmer selling a commoditized good with a price set by the middleman economy is tough, but even that pales in comparison to the loss of autonomy and stress farmers must endure when middlemen don't just set the price but also tell them how to farm.

The pandemic revealed an additional problem with high levels of concentration among these middlemen—concentrated meat processing plants. Because so much meat flows through so few plants, and those plants had been designed to maximize efficiency, that is, to process as much meat as possible using the fewest resources possible, the plants themselves became hotbeds for Covid-19. Employees were forced to work despite insufficient PPE, inadequate ventilation, and an environment that provided workers minimal protection from infection.

During the Trump administration, the Occupational Safety and Health Administration (OSHA) was not particularly active. It did almost nothing in response to the nearly 10,000 Covid-related investigation requests it received during the first six months of the pandemic. But even OSHA decided the meat processors had crossed a line. The only two Covid-related fines it had issued by September 2020 were both imposed on giant meat middlemen.[71] OSHA determined that each had "fail[ed] to provide a workplace free from recognized hazards that can cause death or serious harm."[72] At that time, more than 40,000 workers at meat processing plants had already contracted Covid-19 and more than 200 had died.

The problems at meat processors also affected consumers. Between March and June 2020, the price of poultry, fish, and eggs increased by more than 10 percent.[73] For meat and veal, the increase was over 20 percent, at a time when there was almost no inflation. Making matters worse, these price spikes hit just as Americans were losing their jobs in droves and facing daunting economic uncertainty.[74]

Animal farmers and the animals they raise were also affected. Most farmers, like Laura, now depend on scale to survive. In animal husbandry, this often means raising as many animals as facilities allow and turning them over to be slaughtered as soon as possible. So when processing plants—the key middlemen—had to shut down because of unsafe working conditions, animal farmers had nowhere to send their animals. As a result, despite rising prices and strong consumer demand, many farmers were forced to slaughter their own animals and lay the bodies to waste. An estimated 10 million hens and 10 million pigs were among the animals that farmers "culled" to make room for others.[75] The pandemic revealed the dangers that lurk when too few middlemen stand between hungry consumers and overwhelmed farmers.

The high degree of concentration among meat processors and the scale of other food middlemen helps explain why market forces alone are never going to suffice to address the problems that middlemen pose: The same deep pockets, relationships, control over infrastructure, and other advantages that middlemen accrue to be good at their jobs also allow them to shape the evolution of the markets in which they operate. The largest food companies achieved much of their growth not by developing new products that people wanted to buy, but by buying other companies. In 2018 alone, there were 777 such acquisitions, including twenty-eight where the acquiring company paid more than $1 billion for the company it was buying.[76] This scale gives those food middlemen an outsized influence on the entire food ecosystem.

The rate at which food middlemen have been allowed to gobble up competitors is also further evidence that policy making isn't working the way it is supposed to. In theory, regulators have the power to block any acquisition that would result in a company having too much market power. In practice, any effort to block or condition a merger often results in litigation, giving middlemen the ability to use their money and massive informational advantages to convince a court that the merger should be allowed. Although not the only factor, this helps to explain why the government made so little effort to stop this consolidation of power in the hands of the largest food middlemen.

The way most food travels today captures the two defining fea-tures of today's middleman economy: overly powerful middlemen and long, complex supply chains. It also brings to life the many ripple effects that can flow from the rise of the middleman economy. Today's middlemen don't just connect, they also transform the two ends—shaping how farmers farm and what Americans eat. And in the pro-cess of connecting, they separate, blinding consumers to the people and places behind their food. This helps to explain why there is such a disconnect between the values many Americans hold and the collat-eral consequences of industrial farming, cocoa production, and other modes of food production. At the extreme, the opacity can so impede traceability that deadly bacteria travel farther and infect more victims.

As the next two parts of the book will show, these patterns are pervasive and endemic to today's middleman economy. Many of the dynamics on display here and elsewhere—from increasing returns to scale to the drawbacks of excessive concentration—are familiar and arise in many domains beyond intermediation. Yet, by examining middlemen as such, we can see both why these tendencies are often accentuated when the players involved are middlemen and why the effects can be particularly far-reaching. Combining the familiar yet still important with other, less discussed issues is key to understand-ing how the middleman economy has taken hold and why it matters.

But before digging into the depths of how middlemen became so omnipresent and supply chains so long, it can be helpful to have rea-son to hope. As the next chapter shows, there is a better way.

CHAPTER 2

THE JOY OF GOING
TO THE SOURCE

SEVEN HUNDRED AND eighteen miles east of Laura's farm in Pontiac, Illinois, lies Genesis Farm in Blairstown, New Jersey. The two have a lot in common. Both are predominantly white, predominantly middle class, and predominantly Christian. To varying degrees, both are also rural, with strong agricultural traditions, and both areas attract people who like living in the country. Like Laura, the head farmer at Genesis, Mike, loves to farm and sees farming as his life's work.

Compare Laura's farm to Genesis Farm, however, and the differences couldn't be greater. In contrast to Laura, who grew up farming, Mike didn't know a thing about farming until he was well into his twenties. It was only after Mike had graduated from college, served as a missionary in Kenya, worked for a church in his hometown in Ohio, and provided help to the homeless through a Quaker organization in New York City that he made any foray into farming. He was living in Manhattan and looking to go back overseas when a friend suggested he check out Genesis Farm. The first time he visited, he arrived via bicycle. That he was able to endure the sixty-mile ride, on a very hot summer day, was proof enough of his grit to convince the head farmer at Genesis to offer him a job.

Mike soon moved to the farm and discovered what he had been unable to find in any of his previous jobs: contentment. He liked what he was doing and the way he was living. He was drinking a lot less coffee and eating more vegetables than he ever had before. The pounds were falling off without any effort on his part, and he felt good. He soon got married and followed his wife to England and then back to the United States, but he continued to farm everywhere he went, and he continued to seek out farms that were small and that used techniques that respected the land and surrounding ecosystem. Years later, he got a call inviting him to return to Genesis Farm to replace the head farmer who had given him his start. That was a quarter century ago, and Mike has been there ever since.

Far greater than the differences in how Mike and cousin Laura came to farming, however, are the differences in what "farming" actually entails for each of them. Laura's farm is a high-tech, large-scale operation devoted to maximizing production of corn and soy. Mike and his team use a modest tractor and a lot of manual labor to grow a diverse array of fruits, vegetables, and even flowers. When Mike and his team harvest produce, the food they pick is almost always consumed within days. Laura's dried soy and corn can sit for a year or more. Most of Mike's produce is eaten within twenty miles from his farm; much of Laura's corn and soy, by contrast, travels to the other side of the world.

Mike is able to engage in this type of small-scale farming because Genesis Farm has a very unusual business model. The Community Sponsored Garden at Genesis Farm, the enterprise Mike helps lead, was one of the first "community sponsored agriculture" farms in the country. The model takes a little explaining for anyone not familiar with it. Craig Thompson and Gokcen Coskuner-Balli, researchers at the University of Wisconsin, provide this description: Community supported agriculture (CSA) requires you to "commit several hundred dollars to a local farm," in exchange for which "you will receive a weekly basket of organic produce for a growing season of around six months." Along the way:

> You will have little choice regarding the specific items (and their aesthetic qualities) that make up your basket. Your weekly assortment will be primarily determined by the farmers' planting decisions [and] the exigencies of weather. . . . Invariably, the basket will contain much less of some produce that you prefer[,] too much of some produce you seldom use[, and some you] do not know how to prepare, and may in fact not like. There will also be times when the only way to retrieve your share of a highly coveted crop, such as strawberries, will be to come out to the farm and pick it yourself. Oh yes, there is one other little condition. You will have to make a weekly trek to a designated drop-off point or, in some cases, the farm itself to pick up your basket.[1]

This is just how Genesis Farm works. Topping it off, members then go home with vegetables that they must cook. Mike takes home the same vegetables as his members. He knows how long it takes to transform raw collard greens into a dish that his kids are willing to eat, and how much effort goes into peeling the small potatoes that the farm tends to grow. This is what inspires Mike to quip, "We sell work."

In many ways, CSAs look like a recipe for failure. The standard assumption is that consumers want choice, control, and convenience. Thanks to middlemen, consumers today often enjoy an unprecedented level of all three. CSAs do the opposite. Rather than cater to consumers, they make demands on consumers. They require consumers to pay today for food they will not receive for months, with little say about what they will get or when they get it.

Nonetheless, Genesis Farm is thriving. More than three hundred families now enjoy the seventy-plus varieties of fruits, vegetables, and herbs that the farm grows. The pandemic added to its success, as people gained renewed appreciation of the value of a trusted, local source of food. Even more striking, Genesis Farm is not alone in its success. In 2015, U.S. farmers sold $3 billion in produce directly to consumers. Most of those sales came through farmers' markets and farm stands, but CSAs notched a meaningful $226 million in sales. More than 7,300 farms, spanning the country, sell produce through a CSA.[2] That's up from just two CSA farms in 1986 and about 400 CSA farms in 1993.[3] Although these figures remain tiny relative to total food production, the trend lines show a meaningful rise in direct sales from farmers to consumers, and CSAs appear to be growing in popularity even among the direct options available.[4] This success says something about what today's consumers truly want and value, and why direct exchange may help them to get it. CSAs are thus a great starting point for understanding the benefits of direct exchange.

No need to worry if you have zero interest in ever being part of a CSA. CSAs are used here to illustrate the distinct ecosystem that arises around direct exchange, not as an ideal to which all should aspire. And it's worth bearing in mind that people who join CSAs do not abandon their reliance on middlemen. They do not stop going to the grocery store. Instead, most simply rely on middlemen a little less, and their CSA a little more, to source their daily food.

CONNECTING WITH THE SOURCE

In the neoclassical economics, an exchange is a quid pro quo. Mike gives members vegetables; they give him cash. The exchange makes both better off, but only if each side prefers what it gets to what it gives. This is what economists call "gains from trade." In this view, gains are finite. Every additional dollar that goes to a farmer makes consumers that much poorer. This aptly describes the dynamics at play when farmers like cousin Laura sell their grain. But for a farmer like Mike, this is but one of the many ways that the process of exchange can create value.

The unique structure of a CSA means that members know the

farmer who grows their food, and he knows them. I know this because I am one of Mike's members and his next-door neighbor. Mike's son mows our lawn. His daughter Muriel babysits our young girls. Mike's wife, Kerry, helped to found a local nature-oriented elementary school where many CSA members send their children. Mike and Kerry are part of the community, both through and apart from the CSA. This changes how our family and many other members feel about what we pay and the food we receive.

This is why the story of Mike finding his life's passion and transforming that passion into a livelihood is not irrelevant to the consumer experience at Genesis Farm. Direct exchange often allows for more meaningful work, enabling more people to be growers and makers instead of cogs in a supply chain. But a distinguishing feature of direct exchange is that the satisfaction that comes from meaningful work can extend out from Mike to his customers. I feel good about the money I pay to Genesis Farm each spring not just because of the delicious vegetables I anticipate getting, but because I know the farm and farmers that my money is going to support. I don't want to low-ball any of them.

In a similar vein, when our family savors the Delicata squash and Fairytale eggplants that Mike and his team grow, he hears about it. My kids see him laboring in the fields, and he sees them munching on his blueberries straight from the bushes. Shared joy is not subject to the same rules that govern a quid pro quo. Joy is not a finite good that goes to one person at the expense of another. Sharing enhances the experience, bringing more happiness to all involved. To focus on "gains from trade" is to miss and mischaracterize much of what motivates both Mike and his members.

The CSA structure can also soften setbacks. In 2018, New Jersey had record amounts of rainfall, causing yields to fall across the state. As the growing season came to a close, Mike sent an email to members acknowledging the smallness of their boxes. He explained how the heavy rains caused fennel to rot and onions to perish, and left an entire section of cabbage underwater. He concluded:

If we were part of the typical arrangement . . . , we might be broke. But 30 years ago, a smart group of people decided to try a different arrangement where this risk . . . is spread out among the community. That was and is the thinking behind CSA (Community Supported Agriculture) of which you are a part. . . . While it has been a less than optimal growing season[,] we still have much to be thankful for, like your understanding and continued support.[5]

Other CSA farmers have sent similar messages to their members in bad years, and the response is often the same. Rather than responding with anger or frustration, members respond with empathy and an appreciation of just how hard the farmer tried. Farms and members still suffer when yields are low. There is less food to go around. But the CSA structure reduces the economic hit to farms and can leave all involved grateful for what they received, rather than bitter about what they did not.

Importantly, these dynamics are not specific to the "type" of people who choose to join a CSA. An Oregon woman who gained access to a CSA as part of a study on the eating habits of low-income Americans also found that the structure changed how she experienced her food. "When you saw it was 95° out and you knew [the farmers] are red and they had been out [growing your food], you want to eat every single thing you have and you're not putting it to waste at all."[6] All produce is farmed by someone, somewhere, but the concreteness of a hot, summer day and the awareness that a specific farmer was toiling in a nearby field to grow the food she would soon be eating made her more attuned to the humanity behind her food and more appreciative of that food.

EXCEPTIONAL GOODS

One of the most significant reasons for the success of CSAs may also be the most practical. CSAs often grow amazing fruits and vegetables.

In one survey of CSA members, 93 percent identified the quality of the produce as a reason they joined a CSA, and more than a third said this was the primary reason for their membership.[7] Other surveys have similar results: the freshness of the produce stands out.[8]

There is a good reason CSA produce tastes so fresh: It usually is. Mike requires his apprentices to arrive at work early on pickup days, increasing the amount of produce that can be picked the same day it is taken home. It also helps that CSA members must rejoin for a CSA to survive, but that decision is made only at the end of the growing season. This frees CSA farmers from having to worry about whether a member liked the look of a particular eggplant at pickup, so long as it proves delicious when cooked for supper. It is members' long-term satisfaction that matters.

Farmers who sell their produce through middlemen and via long supply chains don't have these luxuries. They need to grow produce that can be picked long before it is consumed and that can withstand a lot of travel. They also have less ability to prioritize taste over appearance. Prettier produce typically sells better, and sales are what determine revenue. Freshness and taste still matter, but neither can be prioritized in the same way.

EATING BETTER

Many people today want to eat more and different vegetables and live healthier lives. Joining a CSA can help. In one survey, 58 percent of CSA members reported eating more produce as a result of joining their CSA and 74 percent said they ate a greater variety of vegetables as a result.[9] Another study found that Canadians who regularly buy produce from a CSA or farmers' market tend to eat healthier and be healthier than those who don't, even after controlling for factors such as education and income that can also affect eating patterns.[10]

One challenge with many of these studies is what economists call "selection bias." Just showing that people who are CSA members tend to have healthier eating patterns doesn't necessarily mean that joining a CSA leads to healthier eating. It may be that people who like healthy food are more likely to join a CSA.[11] Researchers have found some interesting ways to get around this challenge.

One study minimized the problem of selection bias by comparing the eating habits of CSA members to people who had expressed an interest in joining a CSA, and thus are probably a lot like CSA members, but had not yet joined.[12] The researchers then added a second control group—the CSA members themselves before the growing season had started. Even this study found that getting produce through a CSA had a meaningful impact on what people ate. Six weeks into the CSA season, active CSA members were eating 2.2 more servings of fruits and vegetables each week than people who had merely expressed an interest in CSAs, and active CSA members were also doing a lot more home cooking. CSA members also ate more lettuce and other vegetables after the CSA season started than they themselves had been eating just before the season began.

Another way researchers minimize selection bias is by providing CSA memberships to people not otherwise looking to join a CSA. For example, in one study, a group of public health experts and nutritionists provided subsidized access to a CSA to twenty-five low-income individuals in Portland, Oregon.[13] Participants in the study had to pick up their shares at a designated time and place every Tuesday, for twenty-three weeks. Most paid $5 a week or used public benefits, in exchange for which they got weekly allotments of vegetables valued at around $20. Participants also had opportunities to tour the farm, take cooking classes, and learn about each week's vegetables through bilingual information sheets.

The results make it clear that CSAs are not for everyone. Nearly half of the participants dropped out before the end of the six-month study. Making weekly pickups can be hard for anyone; doing so is even harder if you don't own a car or someone else dictates your work schedule.

For participants who stuck with the CSA, however, the results were significant. Prior to participating in the study, just 17 percent thought they ate sufficient fruits and vegetables. By the end of the six months, that number was 67 percent. A full 78 percent of the study participants reported improved health or health behaviors. And everyone who made it through the six months reported learning new ways to cook vegetables, eating a greater variety of vegetables, and discovering a new vegetable that they liked.

Although every study has limitations, this research shows that how people source their food influences what they put into their bodies. It affects how many vegetables they eat, the variety of vegetables they eat, and how likely they are to cook. These changes came about not by any effort to avoid middlemen, but merely by adding direct exchange to the mix of ways that study participants obtained their food.

PROTECTING THE ENVIRONMENT

As we have learned, dominant modes of growing and distributing food inflict lasting harm on the environment. Too much water is used and polluted, too much land is destroyed, and too much carbon is emitted.[14] Consumers are increasingly attuned to these challenges, which has helped trigger a growing demand for foods that are local and organic.[15] There has also been a proliferation of middlemen, such as Whole Foods, that cater to environmentally conscious consumers. These are helpful but often incomplete steps in addressing the food challenges that have arisen as power shifted from farmers and consumers to the middlemen that connect them.

One challenge is that terms such as "local" and "organic" can be gamed. For example, in some great investigative journalism for the *Tampa Bay Times*, Laura Reiley showed how often the term "local" is abused.[16] She found that restaurants frequently employ very broad definitions of "local," and, even then, still fudge the boundaries. More troubling, many restaurants made outright fallacious claims about the origins of the food they served. She found "wild Alaskan pollock" that turned out to be frozen Chinese pollock, and local "Florida blue crab" that had in fact been shipped from the Indian Ocean. Restaurants also regularly claimed to buy products, from pork to produce, at specific local suppliers that they had never used. In her assessment: "If you eat food, you are being lied to every day." Although her focus was on restaurants, there is little reason to expect greater honesty elsewhere.

"Organic" too is just a label. It is one that a company can use by paying a fee, complying with certain rules, and undergoing an inspection. When consumer and farmer are at a distance, this type of third-party verification can be useful. It provides consumers with information they can trust about how their food is grown.[17] But the quality of that information is limited. A farm can qualify as "organic" and still employ a lot of techniques that don't live up to the spirit that label is meant to connote. Large middlemen and industrial farmers are getting increasingly adept at maximizing yields in ways that can be detrimental long-term while remaining just within the bounds of what the rules allow. For example, a 2018 study found that when the environmental impact of farming is measured against the amount of food produced, organic production techniques are not actually any better for the environment than other modes of farming because yields are so much lower.[18] Michael Pollan raised similar concerns fifteen years ago when he examined the many parallels between large, "organic" growers and other large-scale commercial farms. As he rightly summed up: Organic is "just an imperfect substitute for direct observation of how food is produced."[19]

CSAs provide that opportunity for direct observation and engagement. In a CSA, members don't need some third party to tell them how their food is grown. They can see it for themselves. This helps to explain why less than a quarter of CSA farms are certified organic even though 85 percent use exclusively organic techniques. CSA members can directly observe the clarity of the water in an adjoining river, the abundance of birds chirping and bees buzzing, and numerous other subtle signals of ecological balance. If a member still has questions, she can ask. The same goes for "localness." With a CSA, members know exactly where their food is grown.

Genesis Farm is among the CSAs that grow all of their produce in a way that qualifies as organic, but that is not enough for Mike. He believes that caring for the land also requires crop diversity, small fields to limit erosion, and constant adjustments to minimize impact. As he explains, when farms get too big, "you cannot really pay attention to the detail," and to be a good steward of the land, you have to pay attention to the details. The CSA structure affords Mike and his team the ability to do just that, and his members see the results.

CONNECTIONS AND COMMUNITY

Kate Munning, a baker, writer, mother, and fellow member of Genesis Farm, finds that: "Visiting our CSA isn't just about the food—it's about the people, too. My kids find playmates, I find like-minded cooks and meet folks I didn't even know were my neighbors." For Deborah Debord, a longtime member of a CSA in Colorado, "Bonking heads with other CSA members, exchanging ideas about what to do with the veg, hearing about lives quite different from our own, and watching children grow prove as nourishing as the produce."[20]

Genesis Farm and many CSAs actively cultivate a sense of community among members. They host events with music, food, and activities, allowing members to get to know each other and those who work on the farm. CSA membership can also trigger other types of community building. Some CSA members team up with neighbors to take turns picking up food from the farm. Others share their bounty with neighbors not in the CSA. Again, these effects are not just a byproduct of CSA members being somehow unique. Researchers have found similar community-building gestures when they provided CSA memberships to people who do not fit any particular mold.[21]

These types of opportunities are in short supply today. A growing number of public health officials see loneliness as the next major public health crisis, alongside obesity. Both can lead to more illness and shorter life spans. At its best, direct exchange can help counteract this challenge, providing new avenues for connection, enabling the formation of new types of community, and solidifying existing communities, such as those that arise among neighbors.

SHAPING THE NEXT GENERATION

Kate is not alone in finding a CSA to be a great place for her kids to connect with other kids. I have had the same experience at Genesis Farm. There are child-oriented activities at every event they host, story times at the farm, and even a one-week camp each summer. One study found that 95 percent of CSA members with children brought them to the farm at some point, and almost that many said their kids really enjoyed the visit.[22] The majority valued the ability to have their children spend time on a farm and engage in activities like helping to choose which vegetables to take home. A CSA member in Ohio raves that taking her six-year-old granddaughter to the farm has given the young girl a love of vegetables, including roasted radishes and kale, in addition to allowing her to develop "a relationship with the farmers who are growing those vegetables."[23] A lot of people seem to want kids to have some appreciation of where and how food is grown.

That children might not otherwise understand how their food is grown is a recent development. In 1880, possibly 80 percent of Americans worked in agriculture.[24] Today, that figure is below 2 percent. This means far fewer children are naturally exposed to farming in their daily lives. CSAs make this exposure possible and meaningful. When a CSA member takes his child with him for regular pickups, the child sees how seeds planted one day sprout a few weeks later, and eventually yield the produce that he and his parents take home. A child who was disappointed when rain cut short a visit to the playground can hear from the farmer how those same rains helped feed the corn. These experiences that accompany direct exchange can sometimes provide access and understanding not easily purchased in isolation.

GOOD VALUE

Consumers might care about a farmer's well-being, but many also like getting a good deal. The research available suggests that CSAs often provide consumers just that. One study compared the average total cost that members paid for produce from three different CSAs relative to what those members would have paid for the same amount of organic produce at one of three different types of grocers in the area. The study found that consumers in all three CSAs enjoyed meaningful savings, sometimes paying less than half of the cost of the same amount of produce from one of the middlemen in their area.[25] Another three-year study found that the typical CSA member would have paid 37 percent more had he bought the same produce he got from his CSA from a conventional grocery store instead.[26]

The main reason that CSAs and other forms of direct exchange can provide such value is that eliminating middlemen eliminates the need to pay those middlemen. These costs are substantial. An average of just 15 cents of every dollar that consumers spend on store-bought food makes its way to the farmer who grew it. Most CSA farms engage in labor-intensive modes of farming, and missed pickups or unused veggies can reduce the actual value that any CSA member enjoys, so CSAs will never be the cheapest food available.[27] Nonetheless, the ability to bypass middlemen creates real cost savings, particularly when quality is taken into account.

IDIOSYNCRATIC JOYS AND DEEPER
SOURCES OF MEANING

Being a CSA member has been more challenging than our family expected. It requires planning, late-night stops at the farm with the kids asleep in the backseat, and chopping and cooking when we feel pressed for time. I also feel guilty when we fail to make a pickup or use our full allotment. Because we don't live near the farm full-time, we partake in only a fraction of the events that the CSA sponsors, which adds to the guilt. We have learned how to plan, share, and make more use of the food we receive, but we are not model CSA members, and I doubt we ever will be.

But buying produce this way has also brought unexpected joys. I will never forget the first time we pulled into the deserted farm late on a Friday. As I opened the unlocked door, I found bins of leeks, lettuces, and garlic scapes, and a chalkboard with a handwritten list of what I was to gather from those bins. Sometime between weighing out the new potatoes and trying to figure out how adventurous to be with the greens, I decided to drag my tired husband from the car to join me. He rolled his eyes at first, but soon he too was smiling. The smell intoxicated us both. The cool, damp air hung heavy with a pungent mix of vegetables and dirt. New York City could not have felt farther away. Although I have found many other joys from being part of the CSA at Genesis Farm, the smell of that air remains reason enough for our membership.

Gail, a member of a Massachusetts CSA, found something even more profound through her experience. As she explains, when you live in the city: "You can sit around and read your newspaper and say it's going to rain, must be good for the farmers, but there's no connection. . . . [I]t may be lousy for farmers, maybe that's the day they need dry fields."[28] Joining a CSA helped her to connect these daily realities in a new way and has given her a new "sense of living." For her, "basic food" helped her to feel more connected to all humans. "I'm a human being now, but everything that I have done, having children, loving, losing, grieving over deaths, learning things, these things have been done for millennia." Joining a CSA unlocked within her a new appreciation of the inherent commonality of the human experience.

For much of human history, direct exchange was the dominant mode of transacting because there was no other choice. In joining a CSA, Gail was returning to a more traditional way of doing things, and thus her life was changing, in a small way, to look more like the lives of others who had come before her. Although her experience was unique to her, research shows she is not alone in finding something "spiritual" in CSA membership.[29]

BEING FORCED TO STRETCH

We started with the question of why anyone would join a CSA given the inconveniences and demands that CSAs impose. The answer suggested thus far is that CSAs offer some exceptional benefits. It is possible, however, that the "work" that Mike sells his customers is not just a cost to be justified but a benefit in itself.

Let's look more closely at one of the biggest challenges of CSA membership: Every member goes home with essentially the same allotment of vegetables, including some pretty unusual ones. A member who adores carrots and only tolerates beets will typically receive the same number of beets and carrots as everyone else. Many CSAs refuse to cater even to common preferences. Mike, for example, doesn't decide what to plant based on what members like. He looks instead at what the local climate and soil enable the farm to grow. If a vegetable can be grown in northern New Jersey, Mike tries to grow it. He gives members the sweet corn and blueberries that he knows they love, but he also gives them garlic greens, bok choy, celeriac, kohlrabi, radishes, and turnips.

At first glance, this is a very odd design. The carrot lover seems far worse off than in a system where he could load up on carrots and avoid those pesky beets. But moving past simplistic economic assumptions to the complexities of human psychology reveals that there may be more to the story.

As Kate Munning, fellow Genesis Farm member, notes: "There were plenty of veggies I thought I didn't like when I first joined our CSA."[30] But, over time, she and her kids learned to love foods that once seemed alien, and she has become a better cook in the process. She feels "proud" of the way the CSA helped her and her family to grow. She is not alone. I had the exact same feeling the first time I used a head of cabbage to make a slaw I actually liked, and when I learned how to incorporate rainbow chard into a pasta dish my kids would eat. The more unusual the vegetable, the more satisfaction I feel when I finally discover a way to transform it into something we enjoy.

The burgeoning literature on happiness provides some insight into the sense of satisfaction that Kate and I and so many other CSA members have found in learning to appreciate new vegetables. Gretchen Rubin, a happiness guru who has spent years studying this research, believes "an atmosphere of growth" is critical to happiness. As she explains, whether it's "learning a new language, collecting stamps, or cooking your way through a Julia Child cookbook," the process of taking on a new challenge and growing to meet that challenge is fundamental to well-being.[31] In giving people new vegetables that take them out of their familiar comfort zone, CSAs may be serving up just the raw ingredients some members need to find more happiness.

Now let's go back to Thompson and Coskuner-Balli, the two sociologists at the University of Wisconsin who wrote the overview of the many burdens that CSAs impose on members. They emphasized those challenges not because they dislike CSAs; just the opposite. After extensive interviews with CSA farmers and members and observational work at CSA events, they found that the very things that make CSA membership so inconvenient also made it meaningful, at least for some. Many members found something "enchanting" and even "magical" about rejecting the convenience of fast-food culture for a more rooted, demanding way of eating and living. This was not the uniform experience by any means. But for some members, some of the time, the demands that CSAs impose are not costs that must be balanced out by benefits, but unexpected opportunities to find greater happiness and meaning.

A common feature of direct exchange is that it often takes us out of our comfort zones. One reason for the growth of the middleman economy is that it provides consumers choice, control, and convenience. We have become so accustomed to enjoying all three, we may not even notice how abundant each is. The decision to go directly to the source will often entail compromising along at least one, and sometimes all three, of these dimensions. Without a middleman bridging the gap between maker and consumer, the parties at each end must expend more of their own efforts to form that bridge, and they may have to compromise relative to what they thought they wanted. But as the growth inspired by CSAs makes clear, sometimes getting out of our comfort zones and investing that extra effort can be a benefit in disguise.

CSAs are an extreme example of direct. They will never be the primary way that food moves from farm to table. But you don't need to belong to a CSA or have any interest in joining one to learn something from them. Precisely because they are so extreme, CSAs bring to life many of the practical advantages of direct exchange, like how prices can go down and quality can go up. They also show how direct exchange can enable something more—facilitating connection, forging community, and enabling meaning-making. Even if available and suited for only a small slice of the population, CSAs help explain why direct exchange is on the rise, and why even more direct exchange may make people happier still.

The difference between the heavily intermediated way that most food moves from farm to table, the direct route that CSAs enable, and the way those two different paths affect what consumers consume, how farmers farm, and the ripple effects on the broader world exemplify many of this book's core ideas. These two extremes also serve as a foundation for exploring the vast territory that lies between these two ends.

PART II

THE RISE OF THE MIDDLEMAN ECONOMY

CHAPTER 3

THE RETAIL BEHEMOTHS

NOT LONG AFTER the birth of our second daughter, we decided to escape city life for Thanksgiving. The whole family, with mother-in-law in tow, hightailed it to my aunt's rural getaway in Sussex County, New Jersey. Waking up amid the trees and watching deer out the window proved to be just the balm we needed as worn-out parents of a newborn. The only challenge was that we had underestimated just how much we relied on all of the gadgets we had at home. So, when the time came to make a food run, I offered to go to Walmart instead of the local grocer. The Walmart wasn't much farther and would enable me to get a baby bouncer and extra pacifiers at the same time I stocked up on yogurt and cereal. It seemed like a win-win.

Hours later, I returned with food, baby goods, and lots of other things I had no intention of buying when I set off. Unbeknownst to me at the time, that shopping experience had been decades in the making. There were behind-the-scenes forces that I came to appreciate only years later. Yet the core promise—the ability to get so many different things, at low prices, in one place not far from where we were staying—is one of the most concrete and pervasive embodiments of today's middleman economy.

CHEAP STUFF

My impression that I was getting good deals on everything from yogurt to hair ties on that chilly November day turns out to have a lot of empirical support.* Numerous studies show that as Walmart gained dominance in the early 1990s, it consistently offered lower prices than competitors. Relative to some regional chains, Walmart could offer savings as great as 21 to 28 percent. When compared to other discounters, such as Kmart and Target, the savings were often smaller, 2 to 4 percent on average, but no less consistent.[1]

One challenge in assessing the impact of Walmart on prices and consumer choice is that Walmart stores spread slowly across the United States. This makes it hard for researchers to disaggregate the impact of Walmart's growth from other developments that could also affect how people shop and how much they pay. Economists prefer "shocks"—quick and unexpected changes that make it easier to infer a causal relationship between a given development and subsequent, observed changes.

One development that was far quicker than Walmart's initial growth was its move into the grocery business. This inspired two economists, Jerry Hausman and Ephraim Leibtag, to study how the early 2000s growth of Walmart and other supercenters, which offer

---

* As the first two chapters reflect, the analysis here toggles between two different ways of understanding the forces shaping the economy and the metrics we should use to assess whether it is functioning well. This part primarily focuses on economic efficiency and frictions and other problems that are cognizable in that frame. This helps to illuminate the gains the middleman economy seemed to unlock and why policy makers were so sanguine about many of these changes as they occurred. After laying out their growth in this chapter and the one that follows, the lens again zooms out to add the many important dynamics this framing can elide.

food alongside other consumer goods, affected food prices. Hausman and Leibtag found that supercenters like Walmart really did offer lower prices than traditional grocery stores. Supercenters often sold the exact same food as traditional grocery stores at prices that were 15 to 25 percent lower.[2] Exactly how much the average consumer saved by shopping at a supercenter varied, from 5 percent for bottled water to up to 50 percent for lettuce, but supercenters had lower prices across virtually all of the food products they surveyed.[3]

Walmart's entry into some foreign markets also occurred quickly. Three economists noticed that in Mexico, for example, the number of foreign supermarkets nearly quadrupled—from 365 to 1,335—in a span of just twelve years, from 2002 to 2014.[4] An array of outlets controlled by Walmart de México were the biggest foreign entrants during this time. This is just the type of "shock" that allows researchers to better assess the impact of large middlemen on the well-being of ordinary consumers. The economists were also able to marshal a rich array of data, including confidential bar-code information about the pricing of particular items at different retail outlets each month and detailed information about spending by households within particular stores.

The study's most important finding was that, on average, just about everybody—in their roles as consumers—seemed to benefit from the arrival of Walmart and other massive, foreign middlemen. When a foreign supermarket set up shop in an area, the measurable welfare (a proxy economists use for quality of life) of the average consumer increased by 6 percent. This happened because the foreign chains offered lower prices, new varieties, and additional amenities.

Compared to local stores, the foreign middlemen typically offered prices that were 12 percent lower for the same products. Moreover, the competition from these new stores caused existing supermarkets to lower their prices. As a result, even Mexican consumers who never stepped foot into a foreign grocery store found themselves paying less for groceries.

The entry of Walmart into Mexico also gave consumers more choices. Different people like different things. When my older daughter turned three, she loved yellow. She took far greater joy from her new yellow tutu and sunshine yellow birthday cake than she would have had they been any other color. Choice can make people happier by allowing them to buy goods better suited to their tastes. The researchers found that Walmart and its kin offered Mexican consumers five times as many options as the local alternatives. Richer households shifted more spending to foreign chains and spent more, so they enjoyed more of the upside. Nonetheless, the introduction of giant middlemen in the form of foreign chains seemed to broadly benefit Mexican consumers in the area.

There is also evidence that Walmart helps U.S. consumers pay less, no matter where they shop, by forcing competitors to lower prices and pushing suppliers to find ways to produce goods more cheaply. According to a study by Global Insight, the proliferation of Walmart stores between 1985 and 2006 led to a cumulative 3 percent decline in overall consumer prices. If accurate, this would translate into $287 billion in savings—an average of $2,500 per household—for U.S. consumers in 2006 alone.[5] Again, these gains go disproportionately to the rich who spend more. Moreover, the study was commissioned by Walmart and its findings have not been replicated and so must be taken with a very large grain of salt. Nonetheless, the empirical evidence available does suggest that consumers pay less for goods at Walmart and because of Walmart. Going back to its roots to understand the rise of Walmart is thus a great place to start the investigation into the rise of giant middlemen.

THE BEGINNING

In 1962, on a hot summer day in Rogers, Arkansas, Sam Walton opened the doors of the world's first Walmart store. Sam, then forty-four, had run a number of five-and-dime franchises, but he saw "discounting" as the wave of the future and Walmart was his effort to be part of that wave.

Sam himself concedes, "that first Wal-Mart in Rogers wasn't all that great."[6] As an early manager at the store explained: "[W]e had no real replenishment system[,] . . . no established distributors. No credit." And "everything was just piled up on tables, with no rhyme or reason."[7] Nor was the decision to locate in a small town in Arkansas part of any strategic plan. When Sam first got into retail, he envisioned running a big-city department store. He changed course only because his wife, Helen, refused to live in a town of more than 10,000.

Yet the decision to focus on small towns ended up serving Sam well. As Sam came to recognize, people living in small towns had fewer choices than their big-city peers, so they had more to gain from a new Walmart. He later opined that "the first big lesson we learned was that there was much, much more business out there in small-town America than anyone, including me, had ever dreamed of."[8] Through a combination of luck, drive, and some other dynamics, he addressed each of the other challenges on display at that Rogers store. He developed relationships with banks and distributors and created a department system for organizing goods within the store. He then opened store upon store, allowing him to replicate innovations that worked and tinker with those that didn't.

Two of Sam's great passions were getting good deals and finding new ways to display or price goods to build up customer excitement— "merchandising," as Sam called it.[9] When he opened Walmart #3 in Springdale, Arkansas, for example, Sam decided to make a splash by offering up "truckloads" of cheap antifreeze and "Crest toothpaste at 27 cents a tube." "Well," a colleague recalls, "we had people come from as far as Tulsa to buy toothpaste and antifreeze."[10] Sam instinctively understood the power of a good deal to get people in the door, and how much other shopping they would do once they got there.

By the time he died, Sam Walton had become the wealthiest man in the United States. In the process, he built a juggernaut. Every year, *Fortune* publishes the "Fortune 500," a list of the highest revenue-producing companies in the United States. Collectively, the companies on this list represent two-thirds of the U.S. gross domestic product

(GDP).[11] The list includes household names, such as Ford, AT&T, and Exxon Mobil, along with newcomers, including Apple, Facebook, and Amazon. But one company alone has dominated the Fortune 500 over the last two decades: Walmart. Walmart not only appeared at the number one spot more than any other single company, it was number one a remarkable seventeen times between 2002 and 2021. Shift to *Fortune*'s "Global 500" and the picture barely changes: Walmart is at the top of the list sixteen times over the same two decades.[12]

One reason is that Walmart stores are everywhere. According to Walmart, 91 percent of the U.S. population lives within ten miles of a Walmart.[13] It was not by chance that when I needed pacifiers in rural New Jersey, Walmart seemed like the closest and best option.

Walmart today isn't just *a* middleman. For many suppliers and consumers, it is *the* middleman. Honing in further on two aspects of how Walmart achieved this dominance and where its dominance now plays out—negotiating with suppliers and moving goods from point of creation to Walmart shelves—illuminates how giant middlemen can provide consumers such good deals and some of the ramifications.

PUSHING SUPPLIERS

Demanding the best deal possible from suppliers was part of Walmart's business model from day one. Sam recognized that the less Walmart paid for a good, the better off Walmart and its customers would be. Sam and his team used to fly to New York to meet with supplier after supplier, starting while most New Yorkers were still asleep and continuing after most had finished work for the day. They lived cheap and bargained hard.

The impact of this hard-nosed approach changed as Walmart grew. The bigger it was, the greater the concessions it could and did demand from suppliers. Suppliers the world over know just how big Walmart is. They know that if they want to sell goods to American consumers, there is no better way to do so than to supply goods to

Walmart. Walmart knows this too. Walmart buyers no longer travel to meet suppliers; suppliers now come to them. Most have permanent outposts in Bentonville, Arkansas, where Walmart still maintains its headquarters.

One way to assess just how much leverage Walmart has over its suppliers and how it exercises that leverage is by looking at the impact of selling goods to Walmart on a supplier's bottom line. Gib Carey of Bain & Company did such an analysis. As a consultant, Carey had worked with a lot of clients who negotiated opposite Walmart, and he saw firsthand just how hard a bargain Walmart drove. As he explains it: "If you walk into that annual negotiation with them and you . . . haven't developed new innovations [or] a better delivery system, . . . all you can do . . . is say, 'We got $1.45 per item last year. We'll take $1.40 this year.'"[14]

Carey decided to assess empirically how companies fared when they sold a lot of goods to Walmart. Based on an examination of thirty-eight public companies, each of which earned more than 10 percent of its revenue from Walmart, he found that the more goods that a company sold to Walmart, the lower its operating margins.[15] He separately found that companies that sold a lot of goods to Walmart—meaning that at least a quarter of their revenue came from Walmart—had profit margins roughly *half* the size of otherwise similar companies for which Walmart was not a major customer.[16] Carey could only assess correlation, not causation, but these findings are consistent with the notion that Walmart so squeezes its suppliers that it makes a significant dent in their profitability.

Further evidence of the major price concessions Walmart demands from its suppliers comes from Charles Fishman, author of *The Wal-Mart Effect*. Fishman found that of the ten biggest suppliers to Walmart in 1994, four had been forced into bankruptcy and a fifth had been acquired on the verge of failure ten years later.[17] Despite the high volume of sales that Walmart enables, the fact that half of the company's biggest suppliers couldn't survive the decade is an ominous sign of Walmart's willingness to demand debilitating price concessions.

Walmart's persistent demand for price concessions is bad news for suppliers (and, as we will explore, it can force suppliers to make changes in their operations that spread the pain). But it has resulted in meaningfully lower prices for consumers.

FROM POINT OF PRODUCTION TO POINT OF CONSUMPTION

In addition to negotiating hard, Walmart works hard to save money on logistics. Distance is one of the many hurdles that often stands in the way of transacting. Someone on the other side of the world might offer a great product at a cheap price, but if it costs twenty dollars and two weeks to get it to the consumer who wants it, the value proposition changes dramatically. One of the core ways middlemen provide value is by helping to bridge this gap.

It did not take Sam long to realize that every dollar that Walmart can save on moving goods from point A to point B is another dollar that can be shared between the company and its customers. A well-run distribution network can also reduce under- and oversupply issues. According to one estimate, retail stores fail to have what customers want roughly 8 percent of the time, leading to lost sales.[18] Other times, stores overestimate demand and are left with an excess of products that customers don't want and can only be sold at a discount.[19] Walmart's system allows it to minimize both of these mismatches, further enabling it to offer low prices. These types of cost savings are what economists call "efficiencies" and they are supposed to make everyone better off, ensuring resources are well allocated and reducing waste.

Another way Walmart has reduced distribution costs is by developing close working relationships with the same suppliers with whom it negotiates so hard. This idea did not originate with Walmart. Instead, it was one of Walmart's biggest suppliers—Procter & Gamble—that proposed the approach to Walmart. As one of P&G's vice presidents explained, until that time, there was "no sharing of information, no planning together, no systems coordination" between their two companies.[20] By developing closer ties and working together, P&G believed it could help ensure that Walmart stores across the country got the Tide detergent and Olay moisturizer they needed for customers, while reducing the amount of time those products spent on trucks and in warehouses.

P&G recognized that when Walmart enters into a deal with a supplier, it doesn't have to be a zero-sum game. By pooling their different skills and different information and allowing themselves to become more interdependent, they could save money on operations and logistics while still getting the same product to customers. Soon, Walmart was providing P&G an unprecedented degree of information about just how well its products were selling and where, enabling P&G to use that information in deciding how much to produce and out of which plant. Over time, this also led to new modes of distribution, including direct and timely shipments from P&G to the Walmart stores selling its goods, reducing the need for products to spend any time wasting away in warehouses.

This has changed, on the margins, the nature of P&G as a company. It is still the manufacturer of a host of products that people across the world use on a daily basis, but it is also—in small part—a middleman. Rather than simply trying to maximize production output relative to inputs, it now makes decisions about what to produce and where with the further aim of reducing the logistical burdens of getting its products to where they are needed. This enhances the efficiency of the economy as a whole but also creates an interlock between Walmart and P&G, one that enables Walmart to profitably sell P&G products at a lower price than any of its peers.

As Walmart always does when it realizes it has found what seems like a better way to do something, it then replicated the model. It formed similar partnerships with most of its largest suppliers. It also changed how it worked with smaller suppliers, providing them detailed information on how well their products were selling in various stores, so those suppliers too could enhance operations where possible. As *Time* magazine opined: "Walmart has revolutionized the way retail companies manage their supply chains in more ways than one," but its approach to supplier partnerships may be the "most revolutionary."[21]

Alongside these efforts to develop more symbiotic distribution relationships with suppliers, Walmart also makes continual investments in its internal distribution network. This was one of the few areas where Sam, known for his thrift, was consistently willing to spend. He recognized that spending money on IT and other distribution infrastructure could pay for itself in cost savings over time. By the early 1990s, Walmart had a distribution system that was, in Sam's assessment, "the envy certainly of everyone in our industry, and in a lot of others as well."[22]

Walmart decided early on that if it wanted to make sure distribution ran smoothly, it needed to create its own network instead of relying on third parties, as many other retailers do. It soon devised, and then continued to revise, an integrated system of specialized distribution centers, its own fleet of trucks, and sophisticated IT to track goods throughout the system. Sam then went a step further and integrated efforts to expand the number of Walmart stores with plans to grow Walmart's distribution system. He soon found a model that worked: build a distribution center, stock it with 80 to 85 percent of the products available in a typical Walmart, and then build as many new Walmart stores as possible within 350 miles—a day's drive—of that distribution center.

At the end of Sam's reign in 1992, there were twenty such distribution centers, spanning 18 million square feet, across the country. According to Sam, this system allowed Walmart to replenish store inventories in about two days when others took five. This also allowed Walmart to hold distribution costs to about 3 percent, when competitors had to spend "$4^1/_2$ to 5 percent to get those same goods to their stores," and it provided Walmart greater capacity to meet any unexpected changes in demand.[23] As Sam explained: A store manager could "order merchandise Monday night and get it Tuesday night. Nobody else in the business can deliver like that."[24]

In the decades since, Walmart has continued to replicate this core formula for success, seeking ever more innovative and efficient ways of moving goods across time and space. By 2019, just in the United States, Walmart had 173 distribution centers with 125.8 million square feet of total space, and it employed 8,000 truck drivers who logged nearly 740 million miles per year.[25] This type of infrastructure enables Walmart to move goods exactly where it needs them to be, when it wants them there, and it was unprecedented before Walmart.

A NEW BASELINE

In addition to low prices, Walmart also offers consumers a lot of conveniences. Does a Walmart shopper need to worry about parking? No, Walmart ensures it is free and plentiful. What if she needs bananas along with that laundry detergent? Walmart has her covered. A new showerhead? Walmart has an array of options. Pointy hats, toy-car party favors, and a blue table covering for her son's birthday party? Check, check, and check. Put it all together and Walmart offers ordinary Americans a level of choice and convenience that would have been unfathomable even a few generations back.

According to innovation expert Darrell Rigby, "Every 50 years or so, retailing undergoes this kind of disruption." As Rigby explains in *Harvard Business Review*, these shifts do not "eliminate what came before it." Instead, they "reshape the landscape and redefine consumer expectations, often beyond recognition."[26] The rise of Walmart and other superstores was one such transformative disruption.

This may be Sam Walton's most significant legacy. Walmart didn't just offer customers good deals, it taught them to expect good deals. It created a new baseline premised on cheap prices. And as Walmart continued to proliferate and continued to produce more revenue than any other company in the entire world, year after year, that baseline spread. Consumers everywhere changed. They acclimated to a world that regularly includes trips to Walmart or purchases from competitors offering cheaper prices because of Walmart.

AN INCOMPLETE VIEW

The analysis thus far has focused on how Walmart offers such low prices. This is both because this is what Walmart is famous for and because economists often use price as a proxy. They typically assume, for example, that the less a consumer pays for a good, the more money he can spend on other things so the happier he will be. They often also start from an assumption that when one company can provide a good or service at a lower price than a competitor, it is because the company has found a better, more efficient way of doing things.

This is clearly part of how Walmart can offer lower prices. Walmart's investments and innovations around distribution, for example, create real efficiencies, making society better off, at least in the short run. Walmart's capacity to squeeze ever lower prices from its suppliers is more mixed, even as an initial matter. On the one hand, those suppliers could make changes akin to those Walmart has made in distribution, seeking to reduce waste. But as the negotiations get tougher, and the prices lower, more fundamental changes are often required, such as moving production overseas or reducing quality in ways that consumers might not notice.

Moreover, to fully understand the effect of both Walmart's distribution network and its negotiation tactics, we will need to develop a more dynamic understanding of how these advantages affect Walmart, Walmart customers, Walmart suppliers, and others, in future periods. The upcoming chapters will dive deeper and provide a far more mixed account of Walmart's impact. Those chapters reveal that Walmart's low prices come at a cost. Some of the cost is borne by people and places outside the purview of people shopping at Walmart. Other costs come into fruition only in the future, as large middlemen use their advantages to further entrench themselves and suppress alternatives.

As the shift to looking at direct exchange reveals, however, idealistic efforts to disrupt middlemen or go direct without understanding what it is middlemen do, and why they can be so useful, are often doomed to fail. Understanding how Walmart can offer such low prices is key to explaining both how it has become so big and so powerful, and the challenges inherent in efforts to bring about meaningful change. But the benefits revealed here are only the first step in this story.

THE NEXT GENERATION

Walmart superstores may have set a new bar for choice and convenience in the 1990s, but they have quickly been outdone. Today, just having to go to a store and push around a cart can seem burdensome. Instead, many shoppers expect to get everything I acquired at Walmart, and more, without getting off the couch. In lieu of the handful of options I had to choose among at Walmart, they are accustomed to hundreds of options. They further expect to be able to sort through those options, using star ratings and searchable testimonials, to find just what they want. And if that weren't enough, many consumers today expect the goods to arrive within a matter of days, and sometimes hours, seemingly for free.

This radical transformation explains why the next disruptive wave in Rigby's account was "digital retailing." As recently as 2010, just 6.4 percent of all retail sales occurred online. That figure almost tripled, to 16 percent, by 2019.[27] Just as importantly, the trajectory was upward every single year, and the pandemic only accelerated the shift. And just like the superstore trend, this transformation was brought about by one middleman far more than any other: Amazon.

Amazon's online shopping dominance is so great that most consumers don't bother going elsewhere. As recently as 2015, most consumers at least began the process of shopping for a particular item online by using a search engine such as Google. This provided consumers with an array of options from an array of middlemen and makers. By 2018, things had changed. Two separate surveys and an analysis of web traffic found that when consumers searched for a particular product online, they started their searches on Amazon more often than any other single website.[28] A subsequent survey found that the portion of shoppers who start on Amazon is even higher among those who frequently shop online.[29]

Amazon also captures a sizeable portion of all online retail sales. A subcommittee of the U.S. House of Representatives concluded that Amazon likely captures 50 percent or more of the entire e-commerce market, and that it has an even bigger share in some areas. According to the subcommittee's report, in "October 2020, Amazon's stock price was about $3,000, giving it a market valuation of about $1.5 trillion—greater than that of Walmart, Target, Salesforce, IBM, eBay, and Etsy combined."[30] Amazon may not generate as much revenue as Walmart, but investors clearly see it as *the* middleman that will dominate in the years ahead.

There are a lot of similarities between Amazon and Walmart. In the early days, Amazon looked to Walmart as a model, and it even hired away a number of Walmart executives. More recently, talent has flowed in the opposite direction.[31]

Like Walmart, Amazon is prepared to make major investments today in order to devise a better, more efficient way of doing a task in the future. For example, Amazon took a big jump forward in 2012, when it paid $775 million for Kiva Systems, a robotics company. Within two years, Amazon rolled out its "eighth generation" fulfillment center, a new warehouse design that integrates robots alongside humans to more efficiently move and store goods.[32] By 2017, Amazon had more than 100,000 of these robots employed in warehouses around the world.[33] Search "Amazon Robotics," and you can find videos of orange robots speeding around giant warehouses, carrying loads—up to 3,000 pounds—of shelved goods. Remarkably agile,

they can reach far underneath shelves and into other spaces not readily accessible to mere mortals. By designing warehouses to harness
the differing capacities of the humans and robots in their warehouses,
Amazon can do more work, more quickly, and in less space than its
competitors, or even itself a few years earlier.

Also like Walmart, Amazon's scale is key to its success. Massive
scale is what allows Amazon to make expensive investments in companies like Kiva. Scale is what allows Amazon to build on its success, literally building or reconstructing warehouse after warehouse
based on the most recent iteration of its ideal fulfillment center. Scale
also enables Amazon to gather a huge amount of data about what is
working and what could be improved in a given warehouse design,
enabling its teams of experts to ensure that each iteration of the warehouse system is meaningfully more efficient than the last.

There are also some striking similarities in the founders of the
two companies. Jeff Bezos, like Sam Walton, likes to win. And, like
Walton, he knows that making sure customers are satisfied is critical
to staying on top. Amazon is now one of the nation's most trusted institutions. A 2018 survey found that Democrats trust Amazon more
than any other institution in the survey. Republicans had more faith
in the military and police, but still trusted Amazon more than the
presidency, Congress, state or local government, nonprofits, religion,
universities, Google, or other major companies.[34] Another study
placed Amazon behind only the respondent's primary doctor and the
military as trusted to do the right thing "a lot" of the time.[35] This
may help to explain why both companies have managed to become so
popular and so widely used, even as they attracted some critics.

Both Bezos and Walton also launched their businesses on a wave
of disruption that was transforming the retail structures of their day,
and grew by both riding and being the wave. It is not by chance that
two of the most ambitious men of the past century chose to make
their fortunes as middlemen.

In some ways, the line between Amazon and Walmart is blurring. Walmart is investing in and growing its online presence, while

Amazon has acquired brick-and-mortar grocery store Whole Foods and is building its own Amazon-branded convenience stores. Nonetheless, it is Amazon.com that has transformed the very process of shopping for many Americans. A brief look at some of its key features—customer reviews, an integrated marketplace offering wares from third-party sellers, and Amazon Prime—illuminates why Americans turn to Amazon so frequently, how middlemen provide value to consumers, and how Amazon has engineered a new baseline in customer expectations.

REVIEWS

"Information asymmetries," as economists call them, are one of the key obstacles standing between a producer and consumer. Producers know far more than potential customers about the quality and features of the goods they create. Adding to the challenge, producers can profit from over-claiming quality or hiding defects. Middlemen have long played a role helping parties overcome these informational challenges.

One way that middlemen help overcome information challenges is by putting their own reputation on the line. When consumers buy a good at Walmart, for example, they are in part relying on Walmart to ensure that the good doesn't have hidden defects. Amazon takes a different approach. It intentionally avoids taking direct responsibility for the goods sold via its website. It gets away with this by providing consumers another way to get the information they need: reviews from other consumers.

First, Amazon has users rate each product on a simple five-star scale. By aggregating this feedback into a single rating, Amazon provides a coarse but helpful metric for weeding out products that don't perform as promised and identifying those that do. Second, Amazon allows users to provide detailed feedback through commentary and Q&A. Because the biggest challenge consumers often face is trying to choose which, of an array of options, is the best one for them, this feedback is often even more valuable. The usefulness of both of these tools increases as the number of reviews goes up, and Amazon provides far more product reviews than any other source.

This system is far from perfect. The power of Amazon reviews to sell products creates incentives for sellers to game the system; and because Amazon profits from sales, Amazon may not be taking adequate steps to police fake reviews or ferret out counterfeit products.[36] Nonetheless, people have become increasingly adept at using reviews, shaping how people shop and what they buy.[37] Compiling reviews and making them searchable, sortable, and otherwise user-friendly is one way that a middleman like Amazon provides value. If you have ever looked up a product on Amazon while perusing a store in real life, you have some sense of just how useful this information can be.

THIRD-PARTY SELLERS

One reason Amazon is willing to empower consumers with so much information—including information that may cause a consumer to decide not to buy a particular product—is that the consumer is very likely to choose another option sold via Amazon. From a consumer perspective, Amazon's assortment of products is almost limitless. Want an outdoor table? There are more than four hundred pages of options—with reviews!

Amazon can offer so many choices because many of those products are not actually sold by Amazon. Instead, they are offered by third parties that decide what to sell and at what price, and that bear the risk if the good doesn't sell. In other words, Amazon.com is not really one business; it is two distinct businesses that are seamlessly integrated from a consumer perspective: (1) Amazon the retailer, and (2) Amazon the platform. In both instances, Amazon is serving as a middleman, but merging the two together has been central to Amazon's ability to achieve and maintain such dominance in online retailing.

When Amazon went public in 1997, the volume of third-party sales was around $100 million, about 3 percent of the company's sales. By 2018, those sales had reached $160 billion, implying a compound annual growth rate of 52 percent.[38] This far exceeds the 25 percent average annual growth rate for sales by Amazon. As a result, 58 percent of Amazon's sales in 2018 were by third parties, and that figure is continuing to rise.[39]

Amazon is willing to aid the competition in this way because it still gets valuable data and enjoys a hefty profit even when it "loses" a sale to a third party. This comes not only from the fees that sellers pay to sell goods via Amazon but also from a host of ancillary services that third parties often also buy to try to stand out on Amazon.com. These additional services range from prominent placement to "fulfilled by Amazon" status, which means the seller actually has its goods stocked in Amazon's warehouses and delivered to customers by Amazon's growing fleet of trucks.

The management literature has a term for this strategy: coopetition. The idea is that when competitors collaborate, they can collectively achieve better outcomes than either could achieve on its own. According to experts, Amazon Marketplace is the "the epitome of a coopetition-based business model."[40] As we will see, Amazon is by far the greater beneficiary of these arrangements, but some sellers have managed to create very successful businesses selling goods via Amazon.

The way Amazon's retail business and its platform business feed into and support one another illustrates some of the reasons that large middlemen remain so powerful even as the digital era should seemingly make it easier for makers and consumers to connect without a middleman. As Chapter 10 explores in more depth, dominance is often the norm with online platforms. This is because platforms are "two-sided markets," and each side helps attract the other. As the number of sellers on Amazon increases, so does the number of options, making Amazon more attractive to consumers. And the more consumers who shop on Amazon, the more sellers feel obliged to sell on Amazon in order to reach consumers. Add the value of a consolidated, comprehensive delivery infrastructure, and it becomes clear that digitization itself is no threat to large middlemen.

PRIME

Another vital component of Amazon as it now exists is Prime. As retail expert Doug Stephens explains, Prime is the "vital organ beating at the center of Amazon."[41] For consumers, Prime is the "golden key to the entire Amazon kingdom."

A 2019 survey by Consumer Intelligence Research Partners estimated that a whopping 62 percent of U.S. households are Prime members.[42] The survey further found that Prime members spend an average of $1,400 each year on goods and services acquired through Amazon, more than twice the average amount spent by other Amazon shoppers. Once people have the key to the kingdom, they spend a lot more time and money there.

Amazon knew this when it created Prime. When Prime was launched in 2005, consumers paid $79 a year for unlimited two-day shipping on a slew of products. As Amazon explained to its shareholders the next year, Prime is an "effective marketing tool" because it changes how people shop. Once customers pay for Prime, they have more reason to turn to Amazon for each subsequent purchase.

The price has since gone up, but so have the benefits.[43] In 2011, Amazon added "free" access to TV shows and movies. It later added "free" music and other buying options just for Prime members, such as discounts for monthly subscriptions. And its central commitment to fast, "free" shipping just keeps getting faster and more expansive. By 2019, Amazon Prime promised next-day delivery of a million products to more than 70 percent of Americans.[44] For people living in big cities, same-day delivery is common. Although none of this is actually free to consumers, who pay a large annual fee and provide valuable data with each transaction, the zero marginal cost a consumer incurs each time he streams a movie or has another product delivered can make these benefits feel free in the moment.

One of Amazon's most successful Prime offerings, launched in 2015, is "Prime Day," originally conceived as a Black Friday in July. Amazon now complements good deals with exclusive streaming content. *Rolling Stone* raved that Taylor Swift brought "spectacle" at a New York concert that Amazon sponsored and streamed live to Prime members around the globe.[45] Amazon's status as a primary distributor of music ensured Amazon knew exactly who to invite, and made that invitation far more enticing to Swift and other performers. Amazon's 2019 Prime Day was the biggest sales event in the company's history, exceeding that year's Black Friday and Cyber Monday combined.[46]

Just as with Walmart, these benefits are only part of the story. As we will see, the structure of Amazon, and benefits like Prime, may well give it even more leeway to engage in opportunistic behavior. But to jump to critique without acknowledging the benefits and understanding the interplay between the services Amazon provides and the threats that it poses would do little to help policy makers or the rest of us understand where we are and the real challenges and opportunities that lie ahead.

THE GIANTS JUST KEEP GROWING

When the pandemic hit, upending how people lived and businesses functioned, Walmart and Amazon found ways to keep getting goods to customers. Amazon had $90 billion in sales and more than $5 billion in profits during the second quarter of 2020, when the economy was most impacted by the pandemic. That is double its profits from the same period the year before.[47] Walmart registered $138 billion in total sales and netted $6.5 billion in profits during the second quarter of 2020, despite having earned just $3.6 billion during the same period the year before. Both middlemen were "Covid winners," earning far more during and because of Covid than they would have without it.

This meant that while other companies were furloughing and firing workers, Walmart and Amazon were hiring. This helps to explain why they are not only the two biggest retailers in the country, they are also the two biggest employers.[48] By the end of 2020, Amazon employed 1.3 million full- and part-time employees, compared to just 800,000 at the end of 2019. That is a 62 percent increase in just one year. Walmart may be growing more slowly, but with 1.5 million employees in the United States alone, it remains the largest private employer in the country, and it too is increasing its ranks.[49]

This trend of increasing concentration extended throughout retail. The six largest retail middlemen in the United States—Walmart, Amazon, Target, Home Depot, Lowe's, and Costco—accounted for nearly 30 percent of all U.S. retail sales in the second quarter of 2020. Big middlemen were powerful before Covid but increased their dominance because of it.

This is not to say that these companies did not struggle during this period. They did. But despite the many challenges they faced, they were positioned to leverage their relationships with suppliers, their extensive infrastructure and technology, and other vectors of influence to get as many goods as possible into the hands of consumers that wanted them. And, at the end of the day, it was to these behemoths that people chose to turn, again and again, throughout the pandemic.

Walmart, Amazon, and other large middlemen are so omnipresent that it can be easy to forget just how different the world looked back in 1962 when Sam Walton opened his first Walmart. Large middlemen have not always dominated. Today, they do.

CHAPTER 4

HELPING PEOPLE BUY HOMES

I N HIS 1944 State of the Union address, President Franklin Delano
Roosevelt declared a "second Bill of Rights," including "the right of
every family to a decent home."[1] To help make this new right a reality,
FDR shepherded three bills through Congress, significantly expanding the role of the federal government in housing finance and making
it easier for Americans to buy homes of their own.[2] In the process, he
put home ownership on a pedestal not even shared by health care as
something that all Americans should be entitled to enjoy.

At the time FDR was undertaking his landmark efforts, there was
already a cadre of real estate professionals eager to help would-be
homebuyers find the home of their dreams. Thanks in part to his
housing initiatives, their ranks would swell and there soon would be
an array of banking organizations eager to provide the loans most
people need to buy that home. This chapter examines these two complementary groups of middlemen.

A primary reason to examine housing is its centrality to the wealth
and well-being of ordinary Americans. In contrast to the rich, who
hold most of their wealth in financial assets such as stocks and bonds,

housing is the most important source of wealth for middle-class Americans.[3] Just as important, because housing is so central to wealth for all but the ultra-wealthy, housing also plays an important role contributing to the persistent racial wealth gap in the United States. Understanding the housing market is key to understanding who has wealth and how much they must pay to access it.

Another reason to study housing is that it illuminates parts of the middleman economy that have not yet come to light. Real estate agents can feel far more personal and far more small-scale than Amazon or Walmart. While there's something to that, real estate agents also work collectively to promote their common aims through one of the largest, most well-organized, and most effective trade groups in the country. Their success, in helping clients and themselves, shows how networks of middlemen can provide services and gain power in ways akin to large middlemen. And the ways real estate agents accrue and use their power suggest that there is something about intermediation itself that often contributes to outsized influence, and the problems revealed in this book cannot be attributed just to the scale of the largest middlemen.

REAL ESTATE AGENTS

If you've ever looked for a place to live, you know how difficult it can be to find just what you want. Your home is the place you wake up in each morning, return to each evening, and sometimes never leave. A home can be a place to raise a family, entertain friends, or escape from the world. Owning a home accentuates the stakes, making it harder to move and increasing the financial risk.

This helps to explain why so many people seek some professional assistance when buying or selling a home. But it doesn't explain the particular structure of the real estate market in the United States. Unlike the many other professionals involved, such as lawyers, appraisers, and inspectors, real estate agents are not just service providers that earn a fee for services rendered. They operate instead as middlemen, helping to connect while also consistently standing between buyers and sellers. To see how they managed to achieve and maintain this position, and to earn more than any of their counterparts in other countries, a little history is helpful.

THE MULTIPLE-LISTING SERVICE

The centrality of real estate agents in the buying and selling of homes in the United States goes back to the 1800s. At that time, local real estate agents would gather together to exchange information about their listings. These "exchanges" served the interests of both buyer and seller clients by making it more likely that each would find the right match. Over time, these exchanges became formalized through a multiple-listing service, or MLS.

Unlike any of the goods for sale on the shelves of Walmart, most homes are idiosyncratic. They are unique, just like the people who live in them. This makes finding the right "match" important to both buyers and sellers.

For homebuyers, finding the right match entails satisfying objective criteria, such as school district, size, and number of bedrooms. It can involve trade-offs, such as whether to endure a longer commute for an extra bedroom. And often, it is subtle features, such as the view out of the kitchen window, that help make a house feel like a home. For sellers, finding the "right" buyer usually means finding the buyer willing to pay the most or offer the most certainty. Because a home is often a family's biggest asset, the difference between the best buyer and another, slightly less motivated one can have a meaningful impact on the seller's financial well-being.

The MLS makes this matching process far easier for both buyers and sellers. An MLS is a database with detailed information about homes recently sold or for sale in a given area. To understand the importance of the MLS, it's helpful to understand what economists call "network effects." Sometimes a good or service is valuable not because it is superior to alternatives, but because everyone else uses it. As we saw in the last chapter, a lot of sellers choose to sell via Amazon because so many buyers use the platform, and those sellers attract even more buyers. Similarly, for someone selling a home, inclusion in the MLS means getting her home in front of people actively looking to buy a house, increasing the probability of finding the "best" buyer. For buyers, access to the MLS provides an easy way to identify and sort homes for sale in a given area and to get detailed information about each of them. This makes it easier to find the home that best suits her and her family. By bringing together so many buyers and sellers, and aggregating data and making it searchable, the MLS increases the probability of a good match for both sides.

The MLS also helps real estate agents provide better guidance to their clients as they try to figure out what neighborhood they can realistically afford or how to price a home that they are looking to sell. For example, by allowing a real estate agent to readily identify homes that are comparable to the seller's, the agent can help the seller reduce the risk of pricing a home well below what it could fetch or pricing it so high that it languishes on the market for months.

The usefulness of the MLS is particularly striking when we consider what otherwise may have been required to buy or sell a house before the Internet. Buyers could have spent hours perusing classified ads and looking out for "For Sale" signs, and still have only a fraction of the information the MLS can provide. Sellers could have shelled out a lot of money for classified ads and other modes of advertising, without any assurance that they were reaching their target audience. Sellers also would have had to visit a county clerk's office or rely on potentially inaccurate rumors in deciding how to price their home. This helps to explain why more than 90 percent of all real estate sold around 1980 used the MLS. (Although often referred to in the singular, there are actually numerous MLSs, each owned and controlled by local real estate agents in a given area.)

Today, this may not seem like a big deal. Many people are accus-

tomed to going online and searching seemingly freely available listings on websites such as Zillow or Redfin. But these sites actually
piggyback onto and consolidate listings from local MLSs, rather than
offering an alternative to the MLS. The MLS thus continues to be the
backbone of how buyers and sellers connect, even when they don't
realize it.

RELATIONSHIPS AND REPEAT PLAYERS

The MLS is the cornerstone of how real estate agents help their clients
get valuable information and make the connections they need, but it
is far from the only service they provide. Most people also want to see
a house before they buy it. Real estate agents help make this happen.
The buyer's agent and the seller's agent work together to arrange a
time for potential buyers to visit the home, usually without the seller
present, to see how the light flows into the living room or assess just
how much work a fixer upper really needs. One reason this is usually
such a smooth process is that both agents operate within a network
of relationships, professional norms, and shared long-term incentives
that encourage close cooperation.

Real estate agents are what economists call "repeat players." They
must engage with one another again and again. This allows relationships to form and trust to build. It also means that an agent who lies
or fails to return a phone call can expect retribution in the future,
reducing the likelihood of any bad behavior. This, in turn, makes the
search process smoother for their clients.

Cooperation among real estate agents is solidified by a long-
standing practice in the United States of splitting brokerage fees. Typically, only the seller officially pays out of pocket for the services his
real estate agent provides. The selling agent then splits the commission
he receives with the buyer's agent. As a result, it can feel to the buyer
as if using an agent is costless, even though that is far from the case.
For most of the twentieth century and often to this day, sellers pay

5 to 6 percent of the value of the home sold, meaning each agent gets half that amount. This structure, coupled with the fact that agents usually represent a mix of buyers and sellers, cements the economic backbone that encourages cooperation among agents.

Standing behind and helping to shape these arrangements is a trade group, the National Association of Realtors. Nothing about the structure of how real estate agents work, or how they are compensated, can be understood without reference to the NAR. From its 1908 founding, the NAR has had member groups spanning the country. Today, more than 1.5 million real estate agents are members of the NAR. It issued its first "code of ethics" in 1913, setting out just what was expected of its members. The NAR regularly updates this code, promotes licensing and continuing education requirements for real estate agents, and controls access to the local MLS. In these ways, the NAR provides agents an additional incentive to remain and work in accordance with its standards. Much of the time, this benefits their clients. For example, the code addresses conflicts of interests and candor, making it more likely that an agent will provide a prospective client accurate information about how much a home is likely to sell for and the costs of retaining her to help with that process.

There are benefits to all these features of the real estate market, from the MLS to the economic incentives to cooperate and the role of the NAR in facilitating coordination and promoting honesty among real estate agents. But, as subsequent chapters reveal, it also creates a range of mechanisms for entrenching a system that overcompensates real estate agents and discourages innovations that could make homeowners better off.

HOUSING FINANCE

The other middlemen helping Americans buy homes are the banks and other financial intermediaries that provide the mortgages many need to buy a home that costs far more than they have saved up.

Mortgages bridge the gap between a homebuyer's limited savings and her future income, making home ownership far more broadly accessible than it otherwise would be.

The way access to a mortgage can transform a life was at the heart of the holiday classic *It's a Wonderful Life*. George Bailey, the film's hero, played by Jimmy Stewart, helped a lot of the people around him. Yet his landmark achievement was running the local thrift, a bank that specializes in home loans. As a banker, he helped hardworking members of the community buy their own homes, and avoid having to rent an overpriced slum from Bailey's nemesis, Mr. Potter. In the movie, just as in the stereotype of the American Dream, access to a mortgage is the only way for ordinary people to afford a home of their own and owning a home is the key to a life with some comfort and security. Mortgages still play this role, helping people with decent financial prospects but limited savings to buy a home, avoid paying rent, and invest in their community.[4]

DIVERSIFICATION AND EXPERTISE

Prior to the New Deal, home loans were hard to get. Most Americans who owned homes in the nineteenth century owned them outright.[5] Even at the end of that century, the average mortgage had to be fully paid back within five years and homeowners had to put down an average of roughly 40 percent of the value of the home. This put home ownership out of reach for many. Limited access to financing also changed the types of homes available, most of which were small and crowded by today's standards. As housing experts Adam Levitin and Susan Wachter explain: "Without mortgage finance, Americans would live in substantially different and inferior homes."[6]

As late as 1939, individuals held a third of all outstanding mortgage debt.[7] In other words, savers would give money directly to borrowers. One big drawback of this type of direct finance was that if someone defaulted, the individual lender would suffer a meaningful loss. In part because of this, individual lenders tended to stop making new loans during periods of economic distress, such as the Great Depression. Moreover, they rarely loaned money outside their community, exacerbating structural inequities. This was far from an optimal setup.

Banks mitigate these challenges. One way they do this is by pooling a lot of loans together. To oversimplify, imagine there are 1,000 people who have saved $10,000 apiece and another 1,000 people who each need to borrow $10,000. Let's further assume that 5 percent of the borrowers—fifty people—will default and fail to pay back any money, no matter who makes the loan. If individual savers made loans to individual borrowers, one in twenty would lose their entire savings. By contrast, if all of the savers give their money to the bank, and the bank makes the loans and gives each saver an equal share of the return, each person gets back 95 percent of the money they invested. Add interest, and savers can get a respectable rate of return while being exposed to a far smaller downside risk. This is the beauty of diversification, and facilitating diversification is a long-standing role of financial intermediaries, from banks to mutual funds.

Another benefit of using a middleman in this context is experience. Banks can aggregate data from all of the loans they have made in the past to develop and refine processes and standards that make it easier for them to assess who should get a loan and on what terms. By contrast, most individual savers have no idea how to assess the probability that a particular borrower will pay back his loan in a timely way.

Banks are far more than middlemen. Most depositors want ready access to their savings, whereas borrowers want assurance that they can pay a loan back slowly over time. Banks absorb this maturity mismatch and provide a host of other services as well. Moreover, in ways that are profoundly important for understanding the role of banks in the financial system, but less pertinent here, banks are not just intermediaries but creators of money. In extending loans, they don't just move money from savers to borrowers but actually augment the money supply. Nonetheless, for purposes of this analysis, a core function of what banks do is to serve as middlemen, helping to overcome the informational and other challenges that impede the flow of funds from savers to borrowers.

POLICIES GOVERNING HOUSING AND BANKING

Government policies, starting with FDR and continuing to this day, have played a powerful role in facilitating the growth and determining the shape of the mortgage market. One way the government shaped housing finance was by creating an array of institutions, such as Fannie Mae and Freddie Mac, that will buy loans from banks. Tax policy also encourages people to take out home loans through a deduction that often makes borrowing to buy a house cheaper than borrowing for other purposes.

The government also shaped the mortgage market indirectly through its regulation of banks. The Glass-Steagall Act prohibited banks that accepted deposits, such as commercial banks and thrifts, from also engaging in investment banking. Other laws limited the ability of banks and thrifts to open branches. These laws collectively resulted in an environment where there were a lot of small banks that focused on serving their local communities, taking deposits from people in the area and making home loans to those same people. When *It's a Wonderful Life* was released in 1946, many banks really did resemble the thrift run by George Bailey.

For a while, this worked well. The banking system was stable and home ownership went up, from just 44 percent in 1940 to 63 percent in 1970.[8] Yet as well as this system worked in many regards, it also had some meaningful shortcomings. A core challenge arose from the fact that banks use short-term funding, namely deposits, to fund the long-term home loans. Among other difficulties, this exposes banks to significant interest rate risk. Understanding these dynamics and how they contributed to the savings and loan crisis—a debacle that resulted in the failure of more than a thousand thrifts in the 1980s and early 1990s—helps to explain why the rise of securitization, as the next generation of intermediation, seemed like a good thing.

The development at the root of the S&L crisis was the spike in interest rates starting in the late 1970s. As interest rates went up, and higher-returning deposit alternatives—such as money market mutual funds—proliferated, banks and thrifts had to pay a far higher rate of interest to retain deposits, their primary source of funding. Meanwhile, on the asset side of their balance sheet, banks held a lot of home loans with a fixed interest rate that had been set years earlier, when rates were much lower. The net result was that even thrifts that had been prudent in underwriting loans faced serious financial strains. The situation was exacerbated by decisions by Congress and regulators to ease regulations and enable further risk-taking by weak banks and thrifts. The end result was a massive mess that ultimately taxpayers were forced to clean up.[9]

Against this backdrop, securitization seemed like a useful innovation. Securitization allows banks to move long-term loans off of their balance sheets, and thereby reduce their exposure to changing interest rates. Securitization thus seemed to be the key to a safer and more resilient banking system—one that could be trusted with deposits and could continue to make loans when people wanted to buy a home.

FROM RELIANCE ON ONE MIDDLEMAN TO COMPLEX INTERMEDIATION CHAINS

The core idea behind securitization is that it enables specialization. By making so many home loans, banks had become experts in underwriting home loans. But, as the S&L crisis showed, that didn't mean they were in the best position to bear the risks associated with holding those loans. At the same time, there were a lot of other investors looking to diversify their portfolios who had no idea how to originate a loan and no other way to easily invest in housing. Securitization is what allows investors like pension funds to invest in home loans without having to get into the business of underwriting those loans or engaging with borrowers. Although originally a government innovation, and a core way that Fannie Mae and Freddie Mac helped facilitate mortgage lending by banks, private securitization transactions also started to spread.

In theory, this arrangement benefits all involved. Banks can earn fees by doing what they do best: underwriting and servicing home loans. Investors can reduce the risk of their overall portfolio by adding a new type of investment. And, both of those dynamics should help provide would-be homeowners easier access to financing on more favorable terms.[10] There is also some evidence to support this. For example, a 2015 study by two economists at the Federal Reserve Bank of New York shows that banks were far less willing to extend long-term, fixed-interest-rate mortgages in periods when it was hard for banks to securitize those mortgages.[11] This matters because this is precisely the type of mortgage that people with limited savings and modest incomes need in order to buy a home.

Obviously, securitization creates challenges. One is the need to ensure that banks continue to act responsibly when they originate new loans. To address this, banks usually had to make a lot of promises about the steps they had taken when underwriting a loan. If they lied, they could be forced to buy back the loan. Another challenge of separating loan origination and ownership is determining who should work with the borrower, collecting mortgage payments and following up should a homeowner stop paying. This led to the creation of servicers—often affiliates of the bank that originated a loan—that specialized in providing exactly these services.

Another way securitization enables specialization is by divvying up information burdens. Most mortgage-backed security (MBS) structures—the mechanism through which securitization transforms mortgages into investments—issue multiple, different tranches of MBSs. The most senior tranches, typically given AAA ratings by third-party rating agencies, have the most protections and pay the lowest rate of interest. Meanwhile, through detailed waterfall provisions, lower tranches agree to take on more of the credit and interest rate risk embedded in the underlying loans in exchange for a higher rate of return.

In the process of divvying up cash flows from the underlying mortgages, these provisions also put different types of informational burdens on the holders of the different MBS tranches. It was often rational for investors buying AAA-rated MBSs, for example, to do little due diligence, as the protections embedded in the waterfall structures shielded them from being affected by modest changes in the credit quality and default rates of the underlying mortgages. In an influential paper, economists Gary Gorton and George Pennacchi explain how this divvying up of informational burdens is core to how securitization creates value, as investors were willing to pay a premium for instruments that carry so little default risk that they can be easily bought and sold without meaningful due diligence.[12] This is consistent with the general role of middlemen and other intermediaries in mitigating information challenges, but it is being accomplished in a much different way.

The seeming benefits of securitization, aided by government policies and the central role of government-sponsored enterprises in securitization, help to explain why securitization took off after interest rates spiked in the late 1970s and early 1980s. In 1980, just a little over a tenth of mortgages were subsequently securitized; most remained on the balance sheet of the bank that originated the loan. By the mid-2000s, well over half of all mortgages were sold by the originating bank and placed into some type of securitization structure.[13]

Over time, however, banks were not content just replicating what had worked in the past. Instead, they started to devise new types of tranches, leading to much more complicated MBSs. They also started to create second-layer securitization vehicles, whereby MBSs and other securitized assets were packaged into a new securitization vehicle that would issue new instruments, usually called collateralized debt obligations (CDOs). Again, at least on paper, the model worked. So long as the underlying assets were truly diverse, securitization and tranching could produce information-insensitive, AAA-rated instruments out of riskier assets by redistributing the underlying risks to new higher-return, higher-risk, highly information-sensitive instruments.

THE GOOD THAT PRECEDED
THE BAD AND THE UGLY

Securitization is an example of how the same trends that seem to help explain the rise of middlemen, such as efforts to harness the efficiencies that can arise from specialization, often also lead to layers of middlemen. Before securitization, money would flow from depositor to bank to borrower. There was only one node between the two ends. Moreover, that node—the bank—had a long-standing relationship with both sides. The small banks and bankers that dominated the American landscape for most of the twentieth century knew their clients well. Securitization resulted in multiple nodes separating borrowers from savers. This was accentuated by the fact that many of the "savers" in question were not individuals, but instead themselves middlemen—pension funds and other asset managers investing money for the benefit of yet another layer of beneficiaries.

Alongside leading to longer, more complex capital supply chains, securitization also facilitated more cross-border capital flows. In this context, that meant savings from abroad flowing through securitization vehicles and other structures to U.S. homebuyers. The dollar remains the default currency of the world, and dollar-denominated assets are often in high demand. Regulatory requirements and limited domestic investment opportunities in many countries have often accentuated the demand. As a result, in the 2000s, many investors around the world, such as European banks, started buying up AAA-rated MBSs backed by U.S. home loans.

At the time, these developments appeared benign, even helpful. At least in theory, being able to tap into so many new pools of capital should have had the effect of lowering financing costs for Americans looking to buy homes. And there is some evidence that securitization may have played a helpful role bringing the dream of home ownership into the reach of more Americans. As securitization structures spread throughout the 1990s and 2000s, home ownership rates that had been stagnant for decades started to rise again. In 1994, 64 percent of Americans owned their own home, roughly the same proportion as in 1970.[14] A decade later, in 2004, a record-breaking 69 percent of Americans owned their homes.

Minority home ownership, which has always lagged far behind white home ownership in the United States, increased at an even faster rate. In 1994, just 41 percent of Hispanic households and 42 percent of Black households owned their home. By 2004, 50 percent of Hispanic households and 49 percent of Black households lived in residences that they owned.[15] These figures are still far shy of white home ownership rates, but the differential was notably smaller than it had been in the 1980s and 1990s. Progress was being made, and long, complex securitization chains seemed to be feeding that progress.

As is so often the case with the middleman economy, appearances can be deceiving. Banks, aided by the ability to get loans off their balance sheets via securitization, play a crucial role in helping homeowners get the credit most of us need to buy a home. But as the events of 2008 revealed, these intermediation schemes also had embedded deficiencies that inflicted damage far beyond what anyone could have imagined at the time.

CHAPTER 5

THE MIDDLEMEN BEHIND
THE MIDDLEMAN

I F YOU STROLL into your local Walmart looking for a baby gift, you will likely find a stuffed Peek-A-Boo book, a colorful Tummy Time mat, and other Sassy products. Sassy is one of three major brands owned by Crown Crafts, a leading supplier of baby and toddler goods. The company sells 24 million baby bibs each year; that is more than six bibs for each baby born in the United States. It also sells infant and toddler bedding, swaddles, burp cloths, hooded bath towels, changing pads, sippy cups, teethers, room décor, and a whole host of toys.[1] Nonetheless, the company employs just 131 people.[2]

This is possible because even though the company sells millions of products each year to Walmart, Amazon, Target, and other retail middlemen, the company itself does not actually make any of these products. Instead, most of the goods it sells are produced in China, by people not employed by Crown Crafts, in factories neither owned nor controlled by Crown Crafts. Crown Crafts' official physical presence in China is a mere two thousand square feet, the size of a single-family house in the United States.

This wasn't always the case. When Crown Crafts was founded in 1957, and for three decades thereafter, most of the company's manufacturing was done in Georgia. During that time, it grew from being one of many small but profitable textile companies in the area to a leading producer of velvet and bedding.[3] It also acquired Churchill Weavers, a company with even deeper roots. Based in Berea, Kentucky, an area of Appalachia known for its handicrafts, Churchill Weavers was one of the first U.S. companies to produce handwoven textiles on a commercial scale.[4] A "handwoven tradition since 1922" was its motto and its building is now in the National Register of Historic Places.

This began to change around the turn of the century. In 2000, Crown Crafts sold off all of its Georgia manufacturing operations.[5] Then in 2007, Crown Crafts decided to shutter all of Churchill's operations as well. This was the end of U.S. manufacturing for Crown Crafts.[6]

The reasons were plain. As the local head of Churchill acknowledged: "Since global sourcing has become such a big thing, . . . [w]e no longer have the level of business that it would take to sustain our total operation here."[7] But that didn't reduce the pain of the closure. "People are pretty much devastated," she explained. People in the community took pride in Churchill Weavers. Its closure deprived the community of that pride, took away good jobs, and reduced local tax revenue. In her view: "It's a serious loss for Berea and for the state of Kentucky."

Crown Crafts is now a middleman. It is the middleman behind another middleman and in front of yet another. To say that Crown Crafts doesn't *make* anything, in the sense of actually creating the toys, blankets, and bibs that it sells, doesn't mean the company doesn't *do* anything. In coming up with the designs, for example, Crown Crafts plays roles that go beyond just overcoming barriers to transacting. Like many of today's middlemen, it does more than just intermediate. Yet Crown Crafts is also a middleman. It connects the Chinese factories making Sassy-branded products with U.S. retailers.

It is responsible for moving goods from one shore to the other, usually via a massive warehouse in California. It arranges for shipping, it overcomes information asymmetries by taking responsibility for the quality of the goods, and it communicates with both ends about quantities and preferences. Comparing Crown Crafts today to the Crown Crafts of yesteryear, when it manufactured its own goods in company-owned facilities in places like Georgia and Kentucky, makes clear how the company has evolved to become more of a middleman and less of a maker.

The story of Crown Crafts matters because it is not just the story of one company. It is the embodiment of how the economy has changed since Churchill and Crown Crafts were founded more than sixty years ago. It reflects the rise of hyper-specialization and attenuated global supply chains and the role of giant middlemen in contributing to both.

Today, the typical production process involves far more actors, each making a far smaller contribution to the overall production process but doing what they do on a far greater scale. Cotton farms specialize in growing cotton; mills turn that raw cotton into yarn; factories transform that yarn into fabric, and perhaps change its color; other factories sew and decorate, transforming massive spools of fabric into *Frozen*-themed bedding (another Crown Crafts product); and companies like Crown Crafts then import that bedding, moving it from a factory in China to a Walmart in Duluth, Minnesota, where a parent will pay for it and take it home. This is how the middleman economy works. Node upon node, each doing a little—the thing that it can do most cheaply—until the entire notion of making is so disaggregated that no one company, much less person, can be said to be the maker of any finished good.

The ready supply of Sassy-branded Tummy Time mats and playbooks at Walmart stores across the country helps explain why Crown Crafts evolved from a maker of quilts into a middleman that does some design work. Walmart is not just a customer to Crown Crafts but "the" customer. More than four out of every ten products that Crown Crafts sold in the last couple of years, it sold to Walmart. Crown Crafts needs Walmart and, as we have learned, the key to retaining Walmart as a customer is constantly lowering prices.

Early on, the pressure to lower prices may have brought about true value-creating changes, such as the elimination of excessive administrative processes. But Crown Crafts has been selling to Walmart for a long time. There is only so much slack in any company's operations. To keep on selling to Walmart—and Amazon, Crown Crafts' second-biggest customer—Crown Crafts needed to produce large volumes of very cheap products. Abandoning its U.S. manufacturing and relying on other companies in other countries to produce the goods that it sells was critical to making that happen.

The evolution of Crown Crafts from maker to middleman didn't just occur around the same time that Walmart and Amazon were transforming the retail landscape; it was integrally intertwined with that transformation. This chapter shows that giant middlemen and long supply chains—two defining characteristics of the middleman economy—are often two sides of the same coin. Each props up and gains momentum from the other, resulting in even larger middlemen and even longer and more complex supply chains. This is true in the production of goods and in the movement of money. Understanding why and how these trends took hold, and how pervasive they have become, is key to understanding how today's economy really functions.

HOW AND WHERE GOODS ARE MADE

"Specialization" refers to the process of developing the tools and competence needed to be very good at one particular thing. Increased specialization—of a company, person, or body part—often comes at the expense of having the breadth or flexibility of a generalist. We have already seen some examples of specialization in action, including my cousin Laura's farm and the way securitization transformed finance.

Adam Smith explained the power of specialization in his classic work *The Wealth of Nations*, back in 1776. Smith observed that pin factories churn out far more pins per worker than artisans making the same product. The reason: "One man draws out the wire, another straights it, a third cuts it, a fourth points it, a fifth grinds it at the top for receiving the head...."[8] Having each person or machine to do one thing, and to do it very well and very quickly, increases the output relative to the effort, raw ingredients, and other inputs required. This is how specialization increases productivity.

Henry Ford is famous for harnessing the power of specialization when, in 1913, he introduced the world's first moving assembly line. As a vehicle moved along the line, each worker would add one additional component or complete one task, over and over again. The process enabled a group of workers to produce far more cars, far more quickly than the old process of having teams of two to three work collectively to put together a single vehicle.[9] But it also changed the nature of work, often making it less intrinsically rewarding for the laborers involved.

Greater specialization was at the heart of the industrial revolution, and core to how industrialization enhanced productivity. The degree of specialization that was possible, however, and the scale that could accompany it were transformed by a second trend: increased movement of inputs and goods. Today, roughly 60 percent of all goods and services move across country borders.[10]

Economist Richard Baldwin has shown that underneath the broad term "globalization" are at least two different patterns.[11] During the first wave, innovations that reduced the cost of transporting goods across sea and land increasingly enabled consumers in one country to buy goods made in another. In the second wave, starting in the 1980s, significant improvements in information technology and communications enabled a new type of global system—one in which manufacturing doesn't just shift from one country to another but becomes disaggregated across numerous jurisdictions. The easier and cheaper it was to communicate, the more viable it became to rely on a multi-stage, multi-node process for producing a single finished good.

Returning to Ford's plants, as late as the 1950s, three-quarters of all North American auto parts were made in or near Michigan.[12] By 2005, Michigan manufacturers produced just a quarter of North American auto parts. Innovations that eased intermediation reduced the value of proximity, expanded the competitive playing field, and transformed how goods are made.

None of this would have been possible without middlemen. Each time a step was disaggregated from the rest, a gap was created. This in turn necessitated either a new middleman or that a company become, in part, a middleman. This was true in the nineteenth century, as steam engines reduced the cost of traversing physical distances. And it was even more true in the late twentieth century, as information technology made it easier to overcome information asymmetries. As steam and fiber optics eased intermediation, the amount of intermediation in the economy increased.*

The net result: more intermediation, more middlemen, more companies that are—at least in part—middlemen. This is the middleman economy. Lots of middlemen, connecting and separating, and in the process, changing how people live and work the world over.

FINANCE, GLOBAL AND SPECIALIZED

Finance has undergone similar changes, and for similar reasons. As recently as 1980, there were more than 18,000 banks in the United States.[13] Many were small, community organizations like that run by the fictional George Bailey. These banks were middlemen, but they

---

* Government policies, often pushed by middlemen themselves, were also pivotal to these developments. Crown Crafts' ability to move its manufacturing to China, for example, was made far easier as a result of China's admission into the World Trade Organization in 2001.

created a very short chain. Much of the money that flowed through them came from local depositors and went to local borrowers. Empirical evidence supports the notion that community banks make use of relationships and soft information in a way that is fundamentally different from how big banks work. Small business owners, for example, traditionally found it easier to get a loan from their local community bank, in part because those banks trusted that if the business was successful, the bank would gain a long-term customer.[14] Small banks took the long view.

Soon, however, these small banks were overtaken. By 2005, the number of banks had fallen by more than half, to just 6,500. Meanwhile, the largest banks grew in both size and importance. In 1960, the ten largest banks collectively controlled only a fifth of all bank assets.[15] By 2019, just four banks—JPMorgan Chase, Citigroup, Bank of America, and Wells Fargo—controlled half of the bank assets in the United States.[16] This massive increase in bank concentration, resulting in far fewer banks and a handful of giant banks controlling far more of the market, occurred at the same time and helped to facilitate the rise in securitization. Just as small banks specialize in relationships, large banks specialize in data, models, and standardization—the key ingredients enabling more widespread use of securitization and other innovations that result in capital flowing through longer supply chains.

And just as physical supply chains have become global, so too have the supply chains through which capital travels. For example, by 2007, Citigroup had issued more than 30 million credit cards to consumers in 43 countries outside the United States.[17] It also serviced 52 million bank accounts spanning 25 foreign countries and provided financial services to organizations in more than one hundred countries. A middleman like Citigroup is more accurately understood as a network of interconnected and layered middlemen rather than a single entity. It also exemplifies the dangers of treating a corporation—a legal person—as akin to a real person. When trying to assess the relevant supply chain, that is, the twists and turns through which a good or money travels, a single corporation may in fact have numerous middlemen-like layers within it.

Banks and capital flows became increasingly global in other ways as well. The evolution of the German bank IKB is instructive. The bank was founded in 1924. Its headquarters are in Düsseldorf, Germany, a city of just over 600,000 that sits at the intersection of two rivers, the Rhine and the Düssel. For most of its history, IKB took deposits from people living in the surrounding area and used them to make loans to small and midsized German companies.[18] In the 2000s, IKB started to make changes to its strategy. Rather than focusing on making loans, which would help local businesses and families, it started buying the MBSs and CDOs backed by U.S. home loans described in Chapter 4. As the asset side of its balance sheet started to expand with these U.S. dollar-denominated assets, it decided it should also start to finance its activities with liabilities issued in U.S. dollars. So, it started to issue U.S. dollar-denominated commercial paper. The net result was that a German bank was now providing funding to mortgages in the United States—albeit, funding that flowed through at least one and often two securitization structures— and that same German bank was funding itself with commercial paper, which often ended up in money market mutual funds that then offered deposit alternatives to people and institutions in the United States.

Like Crown Crafts, IKB exemplifies a broader trend. Many European banks increased their exposure to dollar-denominated assets on both sides of their balance sheets during this period. Many Asian financial institutions have made similar decisions more recently. In 2020, for example, a Japanese bank that traditionally served rice farmers lost $3.7 billion as a result of holding debt in large U.S. companies such as Hertz that ended up in bankruptcy.[19] Just as in retail supply chains, prices—theoretically reflecting risk-adjusted returns— determine from and to whom money flows as it circles the globe. If those prices are accurate, these mechanisms can help get funding to the places where there are the most opportunities, but in the process, money can travel far from its origins and often travels through increasingly long and complex chains.

Shifting from banking to capital markets reveals similar trends—increasingly long and complex capital supply chains and very powerful, very large middlemen. Traditionally, in the United States, most people who owned stock in public companies owned their shares outright. In 1950, just 6.1 percent of all of the outstanding stock of U.S. public companies was held by institutions, such as pension funds and mutual funds. The rest was held by individuals.[20]

The landscape has since changed dramatically. By 2009, more than half of public company stock was held by institutions, and that figure was 70 percent for the 1,000 largest public companies. Hence, what had been a direct relationship between individual investors and companies was replaced by a structure where middlemen now separate these two ends. This change has brought with it other changes in where power lies. When individuals owned stock in companies outright, they got to vote on who should sit on the board of directors and other matters. When mutual funds and other asset managers own shares on behalf of investors, it is those middlemen that get to exercise the voting and other rights that come along with stock ownership.

Oftentimes, there is not just one middleman, but layers of middlemen in this context too. For example, most pension funds—middlemen that pool resources of retirees and invest it on their behalf—often invest through other middlemen, such as mutual funds, ETFs, venture capital funds, hedge funds, and private equity funds. There are even "funds of funds," which add yet another layer to the chain. At the end of 2016, for example, more than $380 billion of the money invested in private equity funds flowed through fund-of-fund arrangements.[21] So a retiree might have money in a pension fund (middleman #1), which invests some of that money in a fund-of-funds (middleman #2), which in turn allocates that money to various private equity funds (middleman #3), creating three layers between the beneficiary and the company. Adding to the complexity, the money flowing through each of these nodes is mixed with money from other investors and other middlemen, creating a web of interconnections and common exposures.

Concentration is also an issue in this domain. Even though there are roughly eight thousand mutual fund companies, just four—BlackRock, Vanguard, State Street, and Fidelity—managed more than 27 percent of the total assets under management at the end of the first quarter of 2018.[22] Of those, BlackRock is the biggest, managing more than $6 trillion in other people's money.[23] Taking a step back, we can see that in both banking and capital markets, the evolution of finance bears many similarities to the evolution in how goods are made and sold. In both domains, the rise of giant middlemen and long supply chains feed into each other, with specialization, scale, and cross-border flows all playing a helping hand. The result is large and growing middlemen, and increasingly long and complex chains.

And, just as happened in the production and sale of baby goods, the middlemen involved simultaneously connect *and* separate. Investors can reach more and different investment options and borrowers can tap into different pools of capital. But in becoming more reliant on middlemen, savers and borrowers have grown increasingly shrouded from one another. Sophisticated middlemen facilitate the flow of funds, while also erecting new boundaries. As subsequent chapters reveal, this not only laid the groundwork for European and Asian banks to lose money on U.S. investments they never really understood, but it also reduced the resilience of the overall system.

LIVING IN TODAY'S MIDDLEMAN ECONOMY

The primary function of this chapter is to show that the rise of powerful middlemen and complex supply chains is the rule, not the exception. These two, deeply intertwined phenomena are defining features of today's economy. Although often lost amid discussions framed around globalization, scale, or other features, the rise of middlemen and the spread of intermediation have been at the center of changes in how the economy actually works. The next part of this book reveals reasons to be concerned about just how powerful middlemen have become and the degree of complexity embedded in today's supply chains. Yet before moving on, it is helpful to round out the big picture by looking at some of the myriad ways these new economic structures affect us all.

One thing that is clear is that people alive today have far more access to far more goods at far lower prices. According to one estimate, in 1960, the typical American spent roughly 10 percent of his income on clothing and footwear and got an average of 25 pieces in exchange. By 2013, the typical American spent far less—just 3.5 percent of his income—and got far more—about 70 items.[24] Food prices have also gone down dramatically. The share of income U.S. households spend went from 43 percent in 1900 to just 13 percent in 2003. And for most households, that figure is even lower today, even as a growing proportion of what Americans spend on food is spent on restaurants and takeout.[25] These trends are emblematic of the cost savings enjoyed by Americans and other consumers over the last century.

Access to both financing and investments has also gone up. As depicted in Chapter 4, as banks and other mortgage originators that can sell loans to securitization structures spread, more Americans got mortgages and bought homes. Nor has it ever been easier to invest one's savings. Mutual funds and exchange-traded funds provide investors an easy way to build a diversified portfolio of assets. And for investors that instead want to buy and sell individual stocks, the fees have gone down—sometimes to zero—while the speed has increased.[26] And just as with retail shopping, much of this can be done with just a few clicks from one's home computer.

The process of separating production and consumption has also economically benefited workers in the countries that produce the bulk of today's clothing, footwear, and other goods. Nowhere are these trends starker than in China, the place to which Crown Crafts has outsourced so much of its production. Crown Crafts sells finished textiles (bedding and bibs) and plastics (most of its toys). China leads the world in the production and exportation of both. In 2019, China exported $120 billion in textiles. This is more than the entire European Union, India, the United States, and Turkey—the leading textile exporters after China—combined.[27] China also produced 30 percent of the 359 billion tons of plastics produced in 2018.[28] This helps to explain how China's gross domestic product increased at an average rate of 9.5 percent between 1979 and 2018, far outpacing any other country.[29] And the 1,100 percent increase in U.S. imports from China between 1991 and 2007 shows just how much these trends also changed the origins of the goods that Americans now buy.[30]

Yet, people often pay for the cheap prices they now enjoy as consumers in other ways. For example, as Americans bought more clothing, the proportion of those items made in the United States plummeted from roughly 95 percent in 1960 to just 2 percent in 2013. Research shows that competition from China specifically had a meaningful downward impact on the number of U.S. manufacturing jobs, and that it also reduced the wages paid to those Americans who still work in manufacturing.[31] Crown Crafts really is just one example of a much broader trend.

Terri Graham experienced the painful realities of this new world order firsthand. She had been working for L. R. Nelson, a sprinkler company in Peoria, Illinois, for eight years when she was laid off. She was one of eighty factory workers fired in one fell swoop in 2005. David Eglinton, the president of Nelson, was quite frank about the reason: Walmart. According to Eglinton, Walmart executives said they liked the idea of buying goods made in the United States, but "the cost differential is so great that they told us unless we supply them out of China, we couldn't do business."[32] Faced with a choice between losing the company's biggest customer and evolving from a manufacturer to a middleman, the company chose the latter path. It was the same choice Crown Crafts made around the same time. Walmart won; Terri and her fellow workers lost.

The growth of giant financial intermediaries and long capital supply chains can also affect people's daily lives. For example, as standardized metrics displaced relationships and soft knowledge in lending decisions, the lending process changed and so did the relationship between the borrower and those to whom he owed money. This made it harder for borrowers to seek some leeway in the terms of their loan when they hit a bump, even when that flexibility could actually increase the value of the loan. As we will see, this can hurt not only the borrowers, but also the neighborhoods where they live.

Shifting to consider how the rise of the middleman economy affects the texture of people's lives also reveals limitations inherent in the economics-based lens that has dominated policy making in recent decades. To be man number 9 in Adam Smith's pin factory is a different life than being an artisan who makes pins. And as reflected in the incredibly high rate of churn and low levels of employee satisfaction among many at Amazon, working as or for a middleman can be even less rewarding than being pin production specialist number 9.[33] The rise of the middleman economy means more people work as or for middlemen. It means more people become specialists, and fewer are generalists. It means more people adhering to commands not just from a boss but from abroad. It contributes to consumers who are disconnected and uninformed, and who see little freedom to make different choices even when they want to. The nature of work, the nature of consumption, and the structure of society change.

PART III

THE DARK SIDE

CHAPTER 6

WHO DO MIDDLEMEN REALLY SERVE?

ONE THEORY BEHIND the rise of the middleman economy is that it has taken hold primarily because it helps people, at least in their status as consumers and investors. Walmart, for example, is supposed to help people spend less on the goods they want and need. Amazon is supposed to make it easier for people to get what they want, when they want it. Looking at any exchange in isolation, both promises seem to be fulfilled. Walmart does charge low prices; Amazon does deliver a range of goods very quickly with just a click of a button.

Yet people today do not seem to enjoy the extra wealth and leisure that are supposed to follow from saving money and time. Americans today do spend far less on food and clothing than their grandparents spent on the same goods. Yet this has done little to increase financial security for the average American family. An oft-cited 2018 survey by the Federal Reserve Board of Governors found that just 61 percent of Americans could pay for a $400 emergency without going into debt.[1] Another survey found that a mere 41 percent of Americans could cover a $1,000 emergency from savings.[2]

Looking past the amount of cash that families can access quickly to their net worth does not improve the overall outlook. In 1983, the net worth of the family right in the middle, the median American family, was $82,900.[3] Thirty years later, in 2013, after the full benefits of the middleman economy should have been realized, it was barely any higher at $87,800. To be sure, median wealth increased significantly between those two points in time, largely as a by-product of the housing bubble of the mid-2000s, but it was wiped out by the correction in housing prices and the recession brought on by the 2008 crisis.[4] The lifestyle changes necessitated by the pandemic coupled with rising asset prices have again allowed many Americans to increase their net worth, but those gains too may prove to be short-lived and have been far from evenly spread.

In *The Wolf at the Door: The Menace of Economic Insecurity and How to Fight It*, professors Michael Graetz and Ian Shapiro show that economic insecurity has become the norm for the American middle class and is playing a powerful role contributing to political polarization and "the nativism, racism, and tribalism that feed many populist agendas."[5] If Walmart really "saves" the average American family $2,500 each year, as the study conducted by Global Insight—at Walmart's behest—seems to suggest, this evidence of widespread financial fragility raises the question of where those supposed savings are going.[6]

Data from the Pew Research Center also casts doubt on whether all of the conveniences middlemen supposedly enable actually translate into greater leisure and free time. A 2018 study found that 60 percent of adults sometimes feel too busy to enjoy modern life.[7] In another study, more than half of adults said that they are often trying to do two or more things at the same time. Although there are a lot of reasons people can feel busy and some may enjoy multitasking, it is safe to say that intermediation has not left the average American with a sense of spaciousness in their days.

Shifting the focus to the financial sector further indicates that something is amiss. On the one hand, a number of studies have found that a robust financial sector improves economic growth—banks, investment banks, and other middlemen can help the overall economy by helping entrepreneurs and companies access the financing they need to grow. On the other hand, recent studies show that when the financial system becomes too large relative to the economy, the relationship inverts and further growth in the relative size of the financial sector is correlated with a lower rate of subsequent economic growth.[8] The United States long ago passed the optimal point and is among the countries that may well have too large of a financial sector.

In another study, two Harvard economists investigated the exceptionally rapid growth of the financial sector in the United States, which ballooned from just 2.8 percent of the gross domestic product in 1950 to 8.3 percent of the GDP in 2006. They found that much of the growth after 1980 was attributable to just two changes—more household credit and the growth of asset management.[9] Yet whether these changes make society better off or just line the pockets of financial middlemen remains deeply contested. Access to credit, for example, can help families buy homes and pursue valuable educational opportunities, but debt can also be a burden that detracts from well-being and limits personal and professional freedom. Other leading economists have argued that this growth in debt is little more than a mechanism by which the ultra-wealthy funnel excess savings into financial assets that do little to make society better off.[10]

In another influential study, New York University economist Thomas Philippon tried to figure out just how much the process of financial intermediation had improved given the massive changes in information technology. Improvements in IT should make it easier to overcome the informational hurdles that stand between savers and projects, making financial intermediation less costly. But Philippon didn't find a decline in the cost of intermediation. Instead, examining data going all the way back to 1880, he found that despite all of the innovation, the cost of financial intermediation remained remarkably constant, with only modest fluctuations for more than a century.[11]

Because of the many functions the financial system plays, no single study can support broad conclusions about whether financial middlemen increase or detract from economic vibrance. But looking at this work collectively casts doubt on whether the middleman economy is serving up the benefits it is meant to provide. These studies suggest that even if middlemen and intricate intermediation regimes create value, middlemen are also good at ensuring that much of the value they create (and some they don't) finds its way into their own pockets. This part shows that a "middleman economy" isn't just an economy with large middlemen; it is one that too often enriches those middlemen at the expense of the people who they are supposed to serve.

ANOTHER LOOK AT WALMART

When I took my fateful trip to Walmart to stock up for our Thanksgiving weekend, I succeeded in getting groceries and a baby bouncer in one fell swoop. But I also bought a lot of other things I had no intention of buying when I entered the store. Afterward, as I was loading bag upon bag into our station wagon, I felt bewildered. What had I just done?

On my way to the baby bouncers, I had passed through the girls' department. I noticed a purple dress with flowers that my oldest would like for just five dollars. I tossed it in. Then I saw some hair ties and remembered that she had been asking for a braid. So, I tossed them in as well. At first, it seemed fun.

But I didn't stop when the joy receded. On I went, through the girls' department, the baby department, the grocery area, and more. I was often frustrated and unable to find what I wanted. Yet in the process of searching for those items, I kept finding deals that seemed too good to pass up. "Plum Organics baby food for just a $1 a pack," I thought, "yes please. . . . Napkins with turkeys on them? Not normally my style, but they might add some festivity and at $2.99, why not? . . . Oooh, gingerbread houses, and at a far lower price than it would cost if I baked them myself!" And so it went. And went.

Even when I resisted, I often found myself caught in an internal debate: "Maybe I should do some early Christmas shopping? Sierra would love this Play-Doh baking set and I could get some new onesies for Essie. . . . No, Essie can wear hand-me-down onesies and it is way too early to shop for Christmas. . . ."

Shoppers who regularly visit the same Walmart are probably more efficient, and hopefully more discerning, than I was. But the experience of being both overwhelmed by choice and excessively tempted by seemingly good deals was not unique to me, nor was it something that just happened. From the day he opened Walmart, Sam Walton cared about more than low prices. His self-described "absolute passions" was getting customers *excited* about the deals Walmart offered.[12] He knew that when he got people to cross state lines to buy toothpaste and antifreeze when Walmart #3 opened its doors, they weren't going to go home with just toothpaste and antifreeze. They were going to buy a heck of a lot of other things as well. And just as Walmart's ability to squeeze suppliers has increased as it has grown, so too has its capacity to influence consumer behavior. Today's middlemen have more data and are more sophisticated than at any point in the past, and most are not hesitant to use their insights in self-serving ways. Walmart is designed to get you to spend.

Although I didn't consciously notice it at the time, the cart I was pushing around was bigger than the average grocery cart. That extra space subtly enabled me to pile on more than I otherwise might, as I likely underestimated just how much I had already added—until I got to the checkout. Nor was it by chance that I had to wander through the girls' department, where I didn't really need a thing, to reach the baby department, where I did. The massive size and layout make it more likely that customers will make unplanned purchases along with their planned ones. Big signs advertising low prices further enticed me to add things I didn't need but that just seemed like such a good deal.[13] Nor was I paying much attention to the music or the smell of Dunkin' Donuts as I entered, though both may well have added to my sense of comfort and the amount of time and money I spent.

Like most middlemen, Walmart doesn't publicize the tools it uses to get people to come more often, stay longer, and buy more.[14] But Sam Walton knew that consumer perceptions were just as important as actual good deals in getting customers to open their wallets, and that they could be cultivated. A growing body of research reveals that Sam Walton is far from the only middleman to play on consumer biases.

FROM LOSS LEADERS TO DARK PATTERNS

In 2000, economists at the Chicago Booth School of Business decided to examine what drives grocery store pricing.[15] Do grocery stores increase prices for foods in high demand—as traditional economic theory might suggest—or do they lower the prices on those products to lure consumers into their stores? To answer this question, they collected seven years of detailed data from the second-largest grocery chain in the Chicago area. They found that the chain did indeed reduce the price of products when demand was the highest. Chicago-area families enjoyed cheap tuna during Lent, discounted beer over Memorial Day, and cheaper snack crackers for Thanksgiving. The researchers further found that the grocery chain was significantly more likely to advertise these seasonal deals than other items.

Academics knew to undertake this type of study because even economists recognize that people are not perfectly rational and that retailers know how to exploit consumer biases. Psychologists and behavioral economists, like Nobel laureate Daniel Kahneman, study how it is that people actually make decisions. Kahneman and his colleagues have shown that in real life people do not rationally evaluate all of the information that might be relevant to a decision and then choose the option that maximizes their long-term best interests. Instead, they give some factors far more weight than reason would suggest they deserve while ignoring others entirely. Middlemen know this. This is why grocery stores offer great deals on just what consumers most want: They know that in deciding where to shop, many consumers will pay too much attention to where they can get cheap beer for the long weekend, and too little to the prices the store charges on all of the other goods that they will buy once they are there.

Shifting from the physical world to the virtual world might seem to shift the balance of power. After all, it is far easier to leave one website for another than to walk out of a grocery store and drive to another. Yet even here, retail middlemen have learned how to exploit consumer biases. A survey published in *Harvard Business Review*, for example, found that when people shop online—as opposed to in a store—they actually tend to buy more in each transaction. On average, the basket of goods at checkout is 25 percent bigger online than at a physical store.[16] The authors attribute this in part to the capacity of online stores to stock a greater array of products, but also to a trick commonly used to get people to buy more than they otherwise would: offering free shipping, but only if the customer spends above a given threshold. The study struck home for me. I have definitely bought additional items I didn't need, and wasted far more time than I care to admit choosing them, to avoid paying seven dollars to transport the goods I actually wanted.

In 2010, British cognitive scientist Harry Brignull coined the term "dark pattern" to describe a user interface or other web design feature "that's been crafted with great attention to detail, and a solid understanding of human psychology, to trick users into doing things they wouldn't otherwise have done."[17] As he explains: "When you use the web, you don't read every word . . . you skim and make assumptions." Companies know this and "take advantage of it by making a page look like it is saying one thing when in fact it is saying another."[18] Examples of dark patterns include sneaking additional products into a consumer's online basket or default settings that automatically sign a consumer up to a subscription or newsletter unless they affirmatively opt out.

Subsequent work by researchers at Princeton and the University of Chicago shows just how frequently retail middlemen use dark patterns and even falsehoods to manipulate buying behavior. Analyzing roughly 53,000 product pages spanning 11,000 shopping websites, the researchers found more than 1,800 instances of dark patterns. Of those 1,800 instances, "the majority are covert, deceptive, and information hiding in nature." They further found that "many patterns exploit cognitive biases, such as the default and framing effects."[19] Today's middlemen know human biases and weaknesses and systematically design online environments to suppress critical information and otherwise manipulate people into buying more than they otherwise would. As if that were not enough, the researchers also found 234 instances, across 183 different websites, of middlemen engaging in outright deceptive behavior to contort consumer decision making toward their own ends.[20]

If Sam's cheap toothpaste and discounted fish for Lent represent the first generation of how retail middlemen exploited consumer biases, and dark patterns are the second, we are quickly entering a third and potentially far more pernicious sea change: the use of artificial intelligence, including machine learning, and big data to fine-tune environments to better serve the interests of the middlemen that create them.

Between 2016 and 2019, the proportion of retailers making an effort to utilize artificial intelligence increased from less than 5 percent to nearly 30 percent.[21] AI is not just being used online; it is also being used to influence the design of physical stores. Once again, Amazon is taking the lead. Consumers can now walk into an Amazon Go store, pick up what they want, and just walk out the door without interacting with a single person. The ease of the shopping experience is enabled by an app that the customer must download onto her phone. After swiping on her way in, Amazon can track her movements, including what she looks at and what she takes as she leaves. This allows for a smoother shopping experience by eliminating the need to check out, but it also means that Amazon gets a lot more data along with the money it earns on the exchange.

Not to be outdone, Walmart created its own Intelligent Retail Lab. In 2019, it transformed its 50,000-square-foot store in Levittown, New

York, into a testing ground. The store uses, in Walmart's words, "an impressive array of sensors, cameras and processors . . . connected by enough cabling to scale Mt. Everest five times."[22] Walmart recognizes that the more data it has about how people actually shop and how they respond to particular products, the better armed it can be when making decisions about what products to stock, how to display them, and other ways to increase sales. As Professor Joseph Turow explains in his book *The Aisles Have Eyes: How Retailers Track Your Shopping, Strip Your Privacy, and Define Your Power*, Walmart is part of a "new generation of merchants" that "routinely track us, store information about what we buy," experiment regularly, and then use those collective insights to modify and customize shopping experiences to serve their own ends.[23]

Apart from the many privacy issues that arise as middlemen monitor more and more aspects of how people shop and buy, the rapid rise of big data and constantly improving tools for analyzing that data fundamentally change the power balance between those who have data and those who don't. In this game, middlemen prevail over consumers, and it is the biggest middlemen, with the most data, that win, again and again.

Whenever I buy goods from Amazon or Walmart, I provide them data. I reveal to them that a shopper who buys size 3 Pampers also often buys Aveeno baby wash and Cheerios. They then use this information to shape the recommendations they make to other buyers who go in search of diapers. Similarly, Amazon "knew" to suggest a grow-your-own butterfly kit during the early phase of the pandemic in spring 2020 (which I instantly bought) because it was popular with other people who had been stocking up on the same workbooks and glue sticks I had been buying.

The value of data today lies not just in how it is being used by algorithms, but how it also can be harnessed to train algorithms. This is the power of machine learning. One reason Amazon and Walmart can make such targeted recommendations is that they are constantly getting more and more data that can be plugged into algorithms to spit out ever more accurate recommendations—making it more likely that I will in fact hit "buy" on some item I hadn't known to want. Yet they can also go further and institute processes through which those algorithms are themselves revised via a machine learning system, resulting in even more targeted recommendations.[24]

As is so often the case, this is helpful for customers. As explained in an article co-authored by an Amazon computer scientist, "For two decades now, Amazon.com has been building a store for every customer."[25] Each time you log in to Amazon.com: "It's as if you walked into a store and the shelves started rearranging themselves, with what you might want moving to the front." But it also means that customers may well end up buying far more than they otherwise intended. A study by three academics who specialize in marketing suggests that membership in Amazon Prime may accentuate the tendency for regular Amazon customers to overbuy.[26] Amazon Prime purchasers often feel like they are in control, leading them to trust Amazon more—rather than eroding their trust, as happens when consumers sense that they are being tricked to into buying more. Yet, as the authors explain, "This counterintuitively reinforces shoppers' impulsive buying behavior." This suggests that business models such as Prime "have direct positive effects on both the top and the bottom line for the e-retailer employing them," and it "underscores the fact such a relationship model may come at the expense of the shopper."[27] This may help to explain why Amazon Prime members typically spend more than twice as much on Amazon in a given year than non-Prime members.[28]

This can be added to the many reasons that the largest middlemen continue to grow so quickly, despite the growing distrust of companies like Amazon. The more data a middleman has, the better its recommendations, leading more people to shop there, giving them even more data, and allowing them to build an even greater advantage relative to its peers. It also helps to explain why it can be so hard for individual consumers to resist using the largest middlemen, as the data other people provide to that middleman actually do make it better than its competitors, even if they spend more, over time, as a result. Shifting the analysis from how middlemen encourage excess spending to its effects beyond pervasive financial fragility provides further evidence that all is not well.

SOMETIMES MORE ISN'T BETTER

Let's start by going back to food. Americans today spend far less on food than their grandparents, and they buy what they need with far less effort. Walmart and its supercenters have been central to these trends. But now let us consider what we might expect to find if middlemen don't just get consumers better deals but also succeed in getting consumers to use some of the money they supposedly save to buy more. When it comes to food, those additional purchases would show up not just in a smaller bank account but also in expanding waistlines.

An ingenious study by two economists found that the spread of Walmart supercenters did indeed have just this effect.[29] The two economists, Charles Courtemanche and Art Carden, started by gathering detailed data regarding individual health conditions and health behaviors from surveys conducted by state health departments and the Centers for Disease Control and Prevention. They then matched that data with information about the number of Walmart supercenters, Sam's Club stores (also run by Walmart), and other discount stores in a particular county, and they looked at how both health and the proximity of a supercenter change over time.

After controlling for an array of factors that can affect health behaviors, Courtemanche and Carden found that a causal relationship did in fact exist. The opening of a Walmart supercenter or Sam's Club increased the average body-mass index (BMI) of people living in the area and each "additional Supercenter per 100,000 residents increases . . . individuals' probability of being obese by 2.3 percent points."[30] In their assessment, the findings suggest "that the proliferation of Walmart Supercenters explains 10.5 percent of the rise in obesity since the late 1980s."[31]

Courtemanche and Carden's findings are consistent with the idea that at least some of the money that consumers supposedly save because of middlemen like Walmart gets plowed right back into those same middlemen, even when the extra consumption detracts from health and well-being. And, as reflected in research finding that obesity increases the probability of depression, the human impact of excess pounds can go well beyond increased medical expenses.[32]

Similar patterns arise in other domains. Recall, between 1960 and 2013, the number of clothing and footwear items acquired by the typical American in a calendar year increased from 25 to about 70.[33] Globally, clothing production doubled between 2000 and 2014.[34] Middlemen and the global supply chains they enable allow goods to be made more cheaply, but they also seem to result in people buying even more.

This increase in consumption may not seem like a problem. Extra shirts don't have the same deleterious health effects as too many calories. Yet even here, the impact is far from benign. The average American throws away 70 pounds of textiles each year.[35] The Environmental Protection Agency estimates that total textile waste in 2017 was 16.9 million tons, and a mere 15 percent of it was recycled.[36] This is nearly ten times the textile waste Americans generated back in 1960, when they bought fewer clothes and wore them for longer. That's a lot of trash.

It also far from clear that buying so much more stuff makes people happier. Marie Kondo's first book, *The Life-Changing Magic of Tidying Up,* has spent more than 160 weeks on the *New York Times* bestseller list.[37] The book and the TV show it inspired are worldwide phenomena, and a core theme in both is the importance of getting rid of extra stuff. The centerpiece of Kondo's approach is to focus on joy. Does a particular item bring joy? If not, it should be passed along. In practice this has translated into endless stories of people discarding mounds of clothing and other goods.[38]

Gretchen Rubin, the lawyer-turned-writer-turned-happiness-guru, made a similar discovery: When she set out to increase her happiness, few lifestyle changes brought her as much joy as getting rid of things.[39] The rush she got from cleaning out her closets was so great, she soon found herself volunteering to help friends do the same, even when they didn't request it.

No one has scientifically studied what people choose to discard when they use "joy" as a guiding principle to winnow out their wardrobes and kitchen drawers. But if my experience is any guide, things I add to my cart just to hit a minimum spend threshold are often purchases I come to regret.

Excessive purchases not only can come back to haunt the purchaser, they can also test relationships. Research suggests that couples are more likely to divorce when money—as opposed to something else—is their top area of disagreement, and many fights about money grow out of concerns that one spouse is spending too much.[40] For example, Shanalee and Perry, a Texas couple in their fifties, sought professional guidance to help address ongoing disputes about Amazon. Shanalee uses Amazon for everything from groceries to business items, resulting in more than three hundred purchases on Amazon each year. Although she views the items as necessities, Perry estimates that they "waste $10,000 a year on Amazon purchases."[41] They sought outside advice when their fights about Amazon got so bad that their son erupted in tears, worried that the Amazon purchases had left the family without any money.

Although the couple may well have had differences even without Amazon, what we have learned about Amazon helps to explain how Shanalee and Perry might have such disparate views, setting the stage for rough conversations. Because Shanalee uses Amazon so often, Amazon has a lot of data about her purchasing habits that it can use to make well-targeted recommendations. This could include items that may seem like necessities once she knows about them, but which she never would have sought out otherwise. Perry just sees the growing pile of bills and boxes. There is no right or wrong; just an unhappy couple, a distressed son, and probably many other families having similar fights across the country.

Many people already have some sense of the challenges this book helps bring to light. They are already weary of giant middlemen and the power they hold. Consider Terri, who lost her job in Peoria, Illinois, when Walmart told the company to move its manufacturing overseas. She knew firsthand the way Walmart was hurting Americans in their roles as workers. Nonetheless, she told Charles Fishman: "Do I still shop at Wal-mart? Unfortunately, sometimes I have to. They have things cheaper than other people. I can't afford to pay $2 more . . . I hate it. I'm pinned there."[42]

Terri is what researchers call a "conflicted shopper," someone who actively dislikes Walmart but shops there nonetheless. She is not alone. According to one survey, 15 percent of Walmart shoppers in Oklahoma City were conflicted shoppers.[43] They still shopped at Walmart, more than once a week on average, and they spent heavily, but they didn't feel good about it. This type of internal conflict undermines well-being yet it is not one that often gets measured by economists or policy makers.

Amazon can also stir up mixed feelings and can be similarly hard to avoid. Jonathan Hancock expressed the sentiments of many in his generation when he penned his essay "My Love-Hate Relationship with Amazon." Hancock describes the consternation he experiences as he keeps buying things on Amazon and keeps providing it "full consent to use [his] data—to create the unrivalled service it offers, extend its reach, and continue asserting its dominance." Yet he also appreciates that he "can ask Alexa for suggestions about my Dad's birthday present, then get her to buy it, gift wrap it, and deliver it to his door." That level of convenience helps to explain why Amazon has been a "go-to for most of [his] adult life, and . . . a hard habit to break."[44] Survey evidence shows he is not alone. According to a 2019 survey of Americans between the ages of 18 and 34, 44 percent said they'd rather give up sex than quit Amazon for a year; 77 percent would sooner forgo alcohol for a full year than have to stop using

Amazon.[45] What starts as convenience may well become a habit or even an addiction.

ANOTHER LOOK AT REAL ESTATE

Extra pounds and mounds of clothing are two indications of the waste created by today's middleman economy. Returning to real estate provides yet another concrete example: mass mailings in which real estate agents brag about their recent sales. These mailings do not facilitate the sale of actual homes—the homes depicted have already been sold. They do not provide people information about a new service—homeowners know what real estate agents do. Despite being expensive, environmentally damaging, and pervasive in expensive cities, these mailings do nothing to improve social welfare. Their only effect is to make it more likely that on the off chance a homeowner—who has not indicated any intent to sell his home—happens to be contemplating a sale, the agent who sent the mailing, as opposed to some other agent, gets the job.

Research by economists Chang-Tai Hsieh and Enrico Moretti begins to explain how it is real estate agents can spend so much money on wasteful mailings.[46] Their focus is on the distinct way that real estate agents are compensated—with a seller typically paying 5 to 6 percent of the sales price to his agent, and his agent splitting that fee with the buyer's agent. Using data from 282 cities over a decade, from 1980 to 1990, they found that as housing prices, and therefore commissions, went up, so too did the number of real estate agents.

For example, Hsieh and Moretti found that in 1980, the typical real estate agent in Minneapolis sold about seven homes per year and the typical agent in Boston sold six homes per year. A decade later, real estate agents in Minneapolis were continuing to sell an average of 6.6 homes per year, but agents in Boston were selling a mere 3.3 homes per year. Just looking at the 1990 data might suggest that there is something about the homes or clients in Boston that makes the sales process more difficult. But the fact that agents in the two cities had similar levels of productivity in 1980 makes this unlikely.

The reason for the rapid decline in the productivity of Boston-based agents, according to Hsieh and Moretti, is that housing prices in Boston had doubled over the decade, while those in Minneapolis remained steady. And, as housing prices went up, agents continued to earn about 6 percent of the value of the homes sold, increasing the amount they earned on each transaction. Ultimately, however, agents in expensive cities did not actually end up with higher incomes, because the larger commissions enticed more people to become real estate agents, resulting in fewer sales per agent and more waste—like those expensive mailings.

As Hsieh and Moretti explain, "the cost of finding a customer increases with the number of realtors in the market, without necessarily generating additional benefits to the customer."[47] Overall, they found that the typical broker in a low-cost city was four and a half times *more productive* than the typical broker in a high-price city. For the authors, this is a sign that society would be better off if many of the agents in high-price cities like Boston and New York were doing something else with their time.

Looking abroad provides additional evidence that the U.S. system for transferring homes is overly costly and wasteful. According to data compiled by the *Wall Street Journal*, the Internet and other technological advances have meaningfully reduced real estate commissions in most countries. In Canada, for example, total average commissions paid to real estate agents per transaction declined from 4.5 percent in 2002 to 3 percent in 2015. In Sweden, the typical commission went from 5 percent to 1.5 percent over that period. Other countries, such as the United Kingdom and Australia, also have commissions that total 1.5 to 2 percent of the value of the transaction. Nonetheless, according to the same data set, the typical commission in the United States has declined only modestly, from 6 percent in 2002 to 5.5 percent in 2015.

Although most people buy and sell homes only rarely, the aggregate impact of paying too much when they do is significant. Recall, for the typical U.S. family, home equity is their number one source of wealth. The more money homeowners must pay to real estate agents, the less money they receive when they convert their most valuable asset into cash. That fees in the United States are so much higher than anywhere else, despite massive improvements in technology, is a bright red flag. It suggests that alongside the many benefits that real estate agents offer, they have also used their tools to entrench an outdated, overly expensive regime.

Comparing real estate agents in the United States to those abroad also reveals that the two defining features of the middleman economy—overly powerful middlemen and overly long supply chains—are also present in real estate brokerage once we establish the right baseline. First, in terms of power, the problem does not lie with individual agents, who are often entrepreneurial and hardworking; it lies instead at the level of the system, with actors such as the NAR and local MLSs, who exercise significant influence. Second, in terms of the supply chain, the use of two real estate agents when one would probably suffice, and the professional norms that situate both agents as obstacles to direct communication between buyers and sellers, creates a structure that is longer, more complicated, and involves more degrees of separation than is likely optimal. That real estate agents individually can be sympathetic and well-intentioned does not negate the possibility that they are also part of a broken system, one that illustrates how the act of connecting and separating can be a source of outsized influence that gets used to the detriment of the real people on either end.

PERPETUATING INEQUALITY

All homeowners may be harmed by the self-serving tactics of the NAR and the real estate agents it helps to coordinate. But other ways that middlemen shape the real estate market are even more pernicious and, like so many challenges in the United States, far from race neutral.

Financial fragility is a problem for many lower- and middle-class families in the United States today, but it is particularly pronounced for people of color. The Fed study on household finances cited earlier further found that while 8 in 10 white adults said they were doing at least okay financially, only two-thirds of Black and Hispanic adults felt the same. A report from the Brookings Institution shows that in 2016, the net worth of a typical white family was $171,000, almost ten times that of a typical Black family, then at $17,150.[48] A lot of factors contribute to this disparity, but one of the biggest is real estate.[49] The proportion of white households that own their own home is nearly 30 percentage points higher than the proportion of Black households that own their own home, and the disparity is even greater in many areas.[50] Minneapolis, Minnesota, is not only the place where George Floyd lost his life at the hands of police officers, it is also home to one of the greatest racial housing gaps in the country. In 2018, 75 percent of white households in the Minneapolis area owned their own home, while only 25 percent of Black families could say the same.[51]

Accentuating the challenge, Black families tend to buy less expensive homes in different neighborhoods than their white peers. One study found that even after accounting for neighborhood quality and amenities, homes in majority Black neighborhoods are worth 23 percent less, amounting to $156 billion in cumulative losses.[52] That study and others show that many Black Americans live in largely Black neighborhoods, and the value of homes in those neighborhoods is often lower than in predominantly white neighborhoods. This makes it far harder for Black Americans to use home ownership to accrue wealth in the ways that white Americans have long been able to do.

In *Race for Profit: How Banks and the Real Estate Industry Undermined Black Homeownership*, Princeton professor Keeanga-Yamahtta Taylor shows how a complex web of middlemen, including banks and real estate agents, contribute to these dynamics. She explains how they "wielded the magical ability to transform race into profit within a racially bifurcated housing market."[53] They did this in part by entrenching that bifurcation: Real estate agents, for example, perpetuated the notion of "good neighborhoods" and would steer white clients toward those neighborhoods. At the same time, they would often steer Black clients away from those neighborhoods and toward areas where a majority of the current residents were also Black.[54]

The role of middlemen in housing finance accentuates the challenge. One problem historically was redlining, that is, banks refusing to make loans backed by homes in predominantly Black neighborhoods. Another problem is the type of loan products offered to Black borrowers. The tendency for Black borrowers to be pushed into ill-suited loans was on full display in the proliferation of subprime mortgages in the mid-2000s. As NYU sociologist Jacob Faber explains, securitization, and the capacity to produce particularly high-yielding MBSs from subprime loans, increased the demand for subprime loans. Mortgage originators responded by rapidly increasing the issuance of such loans, from just 7.6 percent of the mortgage market in 2003 to 20.1 percent in 2006.[55]

To figure out just who was being given a subprime loan, Faber gathered data from nearly 4 million home loan applications that were denied, approved for a prime loan, or approved for a subprime loan. Faber found that even after controlling for income, Blacks were 3.6 times more likely than whites to have their loan application denied and 2.6 times more likely to be offered a subprime loan rather than a prime loan. He further found that for all people of color, income was *positively* associated with getting a subprime loan—meaning higher-earning borrowers were more likely to get a subprime loan—whereas for white borrowers, a higher income increased their chances of getting a prime loan. Faber recognizes that there could be an array of explanations, including borrower fraud, but one possibility is that "brokers who earned commissions were incentivized to push larger, riskier loans on potential borrowers" and that they were more likely to push those loans on people of color.[56]

Steering, redlining, and many of the other ways middlemen contribute to segregation and inhibit the capacity of Black families to accrue wealth are illegal, at least in theory. But detection and enforcement have always remained far from complete, enabling many of these practices to continue, even if in less conspicuous forms. Middlemen are not the only ones who have exploited inequality and racism for self-gain. But they are among those who have contributed to the structural problems that now exist, and who have benefited in the process.

FLIP SIDES

The deeper point is that many of the benefits that middlemen provide have flip sides. Middlemen know from years of experience that people want beer for Super Bowl Sunday. The good news is that they use this knowledge to stock up on beer so they can satisfy the heightened demand. But middlemen also use this same knowledge to develop pricing and marketing strategies that exploit known weaknesses in human decision making. They draw people into their stores with the promise of cheap beer because they know those customers will also stock up on a lot of other, higher-margin goods once they are there.

It is the very information advantages, infrastructure, and other attributes that make middlemen so useful that also enable them to serve their own ends at the expense of the rest of us. Middlemen don't just happen to be in a position to entice people to buy more or to perpetuate inequity; they have the means to do so *because* of the role they play in helping consumers and suppliers connect. The drawbacks are intertwined with the benefits.

Accentuating the potential conflict between the interests of consumers and middlemen is that middlemen often do not earn an equal cut on all goods that they sell. Instead, middlemen often earn more on some goods than on others, giving them an incentive to use their influence to push people toward options that earn more for the middleman.

These practices are pervasive and the costs, while often small in any individual instance, can really add up. Expanding waistlines and exploding closets may seem trivial relative to the racial wealth gap and financial fragility. Yet recognizing that so many different ills can be traced, at least in part, back to the middleman economy is key to understanding just how much is at stake in decisions about "through whom" to buy and invest.

CHAPTER 7

MIDDLEMEN PERPETUATING THE NEED FOR MIDDLEMEN

I N 2013, LONGTIME real estate veteran Joshua Hunt decided that there had to be a better way to sell houses. He understood just how much waste the current system engendered and how the Internet should empower buyers and sellers, making them less reliant on real estate agents. So he founded Trelora—"Realtor" mixed up—a firm of licensed real estate brokers who would provide fewer services and charge far less than traditional brokers.

Here was the original business model: Home sellers would pay a flat fee—just $6,000, subsequently split between Trelora and the buyer's agent—to have a Trelora agent list the home on the local MLS and provide basic services, such as arranging showings. Trelora similarly promised that buyers would pay just $3,000 when Trelora helped them to buy a house. Making good on this promise required a little more footwork. Recall, buyers do not pay anything out of pocket to use an agent. Instead, that agent is compensated indirectly, when the seller's agent splits the fee he receives from the seller. To enable buyers to save money, Trelora promised those clients a "rebate" for any amount Trelora received from the seller's agent above $3,000. Say a buyer bought a $300,000 home using a Trelora agent and the seller used a traditional agent. At closing, the seller would pay his agent $18,000 (6 percent of $300,000); that agent would pay half ($9,000) to the buyer's agent; and Trelora would then give its buyer client $6,000 (the $9,000 minus $3,000).

In addition to charging less than full-service real estate agents, Trelora decided to function less like a middleman and more like a provider of services. For example, traditionally, real estate agents control all communications between their clients. So, if a potential buyer has a question about the dimensions of a closet, she would have to ask her agent to ask the seller's agent, who would then ask the seller; the answer would flow through the same chain. Trelora tried to change this by creating a platform that allowed buyers and sellers to communicate with each other directly. Trelora further sought to empower potential buyers by publicly disclosing just how much of a purported "purchase price" would go to the middlemen—the two real estate agents—rather than the seller.

This model did appeal to a lot of people, and the firm grew quickly in Colorado and soon expanded to other states. Many users raved about the service and the money they saved. Trelora is not for everyone, but it is precisely the type of innovative, lower-price alternative that should lead to more choice and lower prices for people buying and selling homes.

Yet Trelora already has been forced to make significant changes in its business model and its long-term success remains uncertain. In the early 2000s, dozens of well-funded companies similarly tried to offer lower-cost alternatives to traditional real estate agents. Yet each one ended up failing.[1] The same pattern has been repeated again and again since then. In 2016, the *Washington Post* proclaimed: "Commissions of 6 percent for home sales once were the norm. That's changing." The article featured the story of a seller who saved a lot of money and had a great experience using SoloPro, which had a business model akin to Trelora's.[2] Two years later, SoloPro joined that vast graveyard of failed start-ups that have tried to offer new and better alternatives for homebuying.

As industry expert Brad Inman observed at a conference sponsored by the Federal Trade Commission (FTC) in 2018: "The real estate homebuying and selling process, for me, always seems on the cusp of change. I see something new and I say, this is really going to change it." Yet, time and again, the transformation doesn't take hold. As he sees it, despite all of the start-ups and innovation, despite the Internet, the "real estate ecosystem has not changed much."[3]

This chapter and the next shift the focus from the way middlemen influence individual decision making to systemic challenges. Although sometimes less intuitive, these systemic ripple effects are key to understanding the many and diverse harms that flow from today's middleman economy. Examining the challenges facing Trelora, for example, helps to explain why dominant middlemen and intermediation systems so often outlast their optimal utility, to the detriment of the people who rely on them.

THE DARK SIDE OF CRITICAL INFRASTRUCTURE

Trelora's effort to provide prospective buyers information about how much of the selling price would go to the middlemen, rather than the person selling their home, was useful to those buyers. It is the type of transparency that engenders healthy competition and enables people to make better decisions by allowing them to see that using an agent is not actually free. Precisely because some well-informed homebuyers may realize that they don't want a full-service agent, however, full-service agents in the area recognized this free flow of information as a potential threat to their business model. So they fought back. And rather than competing by offering lower prices or better service, they had another weapon they could pull out: access to the MLS. The local MLS—controlled by local real estate agents in coordination with the NAR—told Hunt, as head of Trelora, that in providing truthful information to homebuyers, he was violating MLS rules, and could be fined or banned from the MLS unless he stopped disclosing this information.

This threat against Trelora was just the latest manifestation of a century-long effort by traditional real estate agents to use their collective control over the MLS to suppress competition and entrench their high-fee compensation regime. Because the MLS remains the backbone of the U.S. real estate market, populating even seemingly independent websites such as Zillow and Redfin, Hunt didn't feel he could ignore the threat. The result: Trelora stopped posting the information. The traditional real estate agents managed to restore opacity and deny prospective buyers ready access to accurate and helpful information.

There are laws that are meant to prevent this type of anticompetitive behavior and to counter excessive concentrations of power. They are known in the United States as antitrust laws. Federal antitrust officials have filed numerous charges accusing local MLSs and the NAR of limiting access to the MLS in order to shield full-service, full-price brokers from meaningful competition and to suppress innovation. In 2001, for example, the FTC alleged that Realcomp, the trade group controlling the MLS around Detroit, had unlawfully "narrowed consumer choice" and "hindered the competitive process," through rules that placed discount brokers at an unfair disadvantage. Two federal courts agreed. The appellate court found the evidence supported the conclusion that control over the MLS gave Realcomp "substantial market power," which it had unlawfully abused through its policies.[4] There is also a long history of private suits alleging anticompetitive behavior by local MLSs and their affiliates, and courts finding that local MLSs have taken unlawful steps to quash competition.[5]

When none of these actions sufficed to bring about lasting change, the Department of Justice—the other leading federal antitrust enforcer—initiated a major investigation and lawsuit. The resulting settlement with the NAR, signed in 2008, was supposed to be the game changer. It stipulated that the NAR policies governing eight hundred

MLSs across the country unlawfully discriminated in ways that were likely to impede innovation and harm consumers, and it required reforms designed to level the playing field. A decade later, government regulators held a conference to assess the impact of that agreement. The experts that gathered acknowledged that the agreement resulted in some improvements for consumers. But the overarching message was that the deal had failed to facilitate true disruption. The core market structure—two middlemen and an excessively high and excessively rigid fee structure—had not changed. Moreover, the NAR and local MLSs continue to use MLS rules and website features in ways that disadvantage innovators, including Trelora.

Though the MLS was once a remarkable advance that benefited home buyers and sellers, it is now a mechanism of entrenchment. It is regularly used by the NAR and member agents to perpetuate a system that places too much money in the hands of real estate agents, and too little in the hands of people selling their homes.

The dynamics that enable real estate agents to impede change that could benefit homeowners are not isolated to real estate. Instead, they arise regularly whenever dominant middlemen create and control critical infrastructure.

The potential for self-serving behavior is often most pernicious when middlemen control what economists call a "two-sided market."[6] As we will later examine in more detail, a two-sided market is one that connects two types of users: a newspaper connects readers with people posting classified advertisements, the MLS connects home-buyers and home sellers, Amazon Marketplace connects sellers and buyers around the world. The economics literature on these markets reveals that there is often just one or a handful of winners in these domains and that even a platform that is far from optimal can remain dominant despite infirmities. This happens because the platform has what users of each type most want: access to users of the other type.

Recall, for example, when people shop online today, the majority start their search at Amazon.com. This means that if you are trying to sell a good, you may choose to sell on Amazon even if you dislike aspects of how the company operates because being on Amazon.com is key to being seen by the majority of buyers. And, as new sellers keep flooding toward Amazon, Amazon manages to have more options than any other platform, increasing the tendency for buyers to go just to Amazon. These dynamics provide leeway to Amazon and other dominant platforms to engage in self-serving behavior without necessarily losing many users as a result.

This was one of the reasons that the U.S. House Judiciary Committee's Subcommittee on Antitrust, Commercial, and Administrative Law launched a major investigation into the market power enjoyed by Amazon and other large digital platforms. According to the subcommittee's report, "Interviews with sellers, as well as documents that Subcommittee staff reviewed, make clear that Amazon has monopoly power over most third-party sellers and many of its suppliers."[7] Like the MLS, Amazon is no longer just one platform of many; it is *the* dominant platform. Sellers don't work with Amazon because they want to; many sell via Amazon because they feel they have no real choice.

Yet the critical infrastructure that has enabled Amazon to amass so much power goes beyond its digital platform. Its extensive physical and logistics infrastructure accentuates this influence. Walmart and Amazon both have state-of-the-art distribution centers throughout the country, fleets of trucks and drivers, and technology that allows them to know precisely what goods are where and how they can be moved quickly and efficiently to where they are needed. The economies of scale enabled by the volume of goods that these companies transport provide real benefits to consumers, but in the process, they also give both companies a massive advantage over other middlemen and direct purveyors of goods. If a maker would have to pay $20.00 to get its good into the hands of a customer in two days because it would have to pay for packaging and a third-party shipping service, while Amazon or Walmart could harness internal infrastructure to get it there just as quickly at a marginal cost of just $1.00, any decision not to go through Amazon or Walmart brings significant additional costs that someone must bear.

Prime accentuates Amazon's ability to leverage its infrastructure to its advantage. A consumer may join Prime just for the free movies and Whole Foods delivery. Once that commitment is made, however, the consumer can enjoy delivery that is not only incredibly fast but seemingly free on everything labeled "Prime." This makes it that much more costly for a consumer to choose to go anywhere else. The high proportion of Amazon shoppers that are Prime members also limits the autonomy of third-party sellers. In theory, a third-party seller can either deliver its goods directly to customers or pay a fee for Amazon to warehouse and deliver the goods. In practice, paying the extra fees to Amazon means that the good also gets "✔ prime" on its listing. Sellers know that roughly 62 percent of American households are already Prime members, and they have no way to offer free overnight shipping themselves, so the great majority opt in. According to the subcommittee report, 85 percent of the top 10,000 third-party sellers on Amazon.com pay for Amazon to store and ship their goods, despite the additional costs and loss of control this entails. This not only means Amazon earns more on these sales, it yet further enhances the value of Prime by increasing the number of products that consumers can get delivered next day at no cost.

This is one of the reasons that the tools Amazon has devised for serving clients, including the exhaustive reviews, extensive physical and technological infrastructure, and Prime membership discussed earlier, add up to far more than the sum of those parts. Reviews attract customers who leave more reviews. Third-party sellers come for those customers, who come more often because of those third-party sellers. More items shipped free and fast because they're labeled "Prime" leads more people to become Prime members, which in turn entices yet more third-party sellers to choose Amazon distribution. And the more sellers that show up, the harder it is for any seller to grab attention, compelling sellers to shell out even more for privileged positioning. . . . These advantages are not just additive, but co-constitutive, feeding into each other at turn after turn.

Although the details vary, the common pattern is a cycle:

Middlemen invest money and effort to build infrastructure that makes it easier for them to connect buyers and sellers. The investments these companies make enable them to provide more convenience, better prices, and other consumer-friendly benefits. But it also means that what consumers must forgo if they opt to go elsewhere is that much greater. This increases the number of consumers that use that middleman, allowing it to make further investments in the size of the moat surrounding its fortress. It is not impossible to displace a dominant middleman or network of middlemen that control critical infrastructure, but it is very hard. The very infrastructure that allows middlemen to be useful thus becomes a means for entrenching their dominance.

THE DARK SIDE OF COOPERATION AND RELATIONSHIPS

The relationships that middlemen cultivate to become good middlemen accentuate this vicious cycle. We learned that real estate agents, for example, are repeat players and most represent both buyers and sellers. When added to the NAR's code of ethics and the distinct compensation structure, this leads to a high degree of cooperation among real estate agents. Most of the time, this benefits buyers, sellers, and agents, making the process smoother for all involved.

But now let's examine the implications of this regime for a start-up like Trelora. Local brokers in the Denver area did not respond warmly to Trelora. According to Hunt, Trelora and its clients have endured bricks thrown through windows and cars being keyed. He has received a barrage of angry messages such as: "You're pissing off every broker that wants to show your house so nobody wants to show your house."[8]

The importance of cooperation in real estate, however, meant that when these agents got angry, they also had ways to get even. By 2018, Hunt had compiled "a list of over 719 brokerages in Denver alone that have flat out said, we will not show Trelora listings." This is bad news for the clients of those other real estate brokers. It means that if the house that is perfect for them is being sold by a Trelora agent, their agent may intentionally avoid showing it to them, and they may end up buying a house that costs a little more or suits them a little less. It is also bad news for sellers who want to work with Trelora. It means that using Trelora's original approach could result in his house being shown to fewer prospective buyers, potentially reducing the price it fetches or delaying when that sale happens.

Eventually, this motivated Trelora to revise a core element of its business model. Trelora started telling "every seller, [that] 40 percent of agents will go out of their way, above and beyond, and push hard not to show or sell your home if you don't offer a 2.8 percent to 3 percent commission." So even though it is far from clear that a buyer's agent is doing anything to deserve that type of fee, Trelora encourages its clients to offer it. A cynic might say that these agents have to be bribed into doing right by their own clients.

Other nontraditional brokerage firms have faced similar challenges. Redfin is among the most successful new real estate platforms. But even Redfin has similar stories of losing business because of efforts by traditional brokers to protect high-fee arrangements. In 2018, the CEO of Redfin held up a client in Seattle as an example. Redfin served as the broker in connection with the sale of the client's home, saving the client more than $100,000 relative to what he would have paid a traditional broker. The client also initially wanted to work with Redfin to buy the new, fancier home. When the time came, however, the seller (or really, the seller's broker) opted to use a "pocket listing"—a way of keeping the listing semiprivate—so the client felt like he had little choice but to abandon Redfin in favor of a traditional real estate agent.[9] The unique nature of homes means that the capacity of agents on one side of the transaction—buy or sell—to steer clients away from parties on the other side who opt to use a discount brokerage creates a real barrier to change.

Inspired by the pervasiveness of these types of stories, Steven

Levitt and Chad Syverson, economists at the University of Chicago, decided to investigate whether they could find any evidence that agents did in fact steer clients away from homes being sold by a discount broker.[10] They analyzed data from three different cities over a twenty-month period and controlled for a range of factors that typically affect a home's attractiveness to buyers, such as location and size. They found that sellers opting for the fixed-fee model common to discount brokers saved money, but they were more likely to be unable to sell their home at a price they found adequate, and, even if they did manage to sell it, they typically had to wait longer than a neighbor selling a similar home using a full-service agent. These types of findings suggest that the difficulties faced by Trelora and Redfin are the norm, not the exception.

And again, these types of dynamics are common when middlemen are involved. Investment banks also sometimes function as middlemen. When companies seek to go public for the first time, for example, investment banks have long played a useful role helping to connect those companies with investors and helping those investors understand why the company's stock may be a good investment. Yet, just as with the real estate market, the fees investment banks charge to help take a company public remained remarkably high and consistently so even as changes in technology should have made it much easier to overcome these hurdles.

For all but the largest IPOs, investment banks normally collect a fee of 5 to 7 percent of the offering, reducing the amount that flows to the company and the value of the company to investors. As middlemen, the investment banks also gain the right to allocate IPO shares, a right that itself is quite valuable when shares are expected to go up in price on the first day of trading.[11] According to one study of the more than 650 traditional IPOs (that is, involving an investment bank as underwriter/middleman) that occurred between 2015 and early 2021, the median company going public incurred costs that totaled 21.9 percent of the funds raised. This figure includes both the actual underwriting fees the company paid to the investment bank and the cost the company indirectly incurred as a result of underpricing its shares in that initial offering. It is this systematic underpricing of IPO shares that makes access to those shares so sought after, and investment banks' ability to allocate them so valuable.[12]

That IPOs underwritten by investment banks remained virtually the only way for private companies to go public until quite recently is particularly striking given just how much the information and institutional environment around IPOs has changed. The Internet and other innovations facilitate the flow of information, making it easier for investors to do their own research and develop their own views regarding a company's value and prospects. In addition, today's public company investors are more sophisticated than in previous generations. Most money flowing into public companies today flows through institutional investors who are better positioned to assess a company's prospects than the retail investors who dominated the landscape when investment banks originally came to play such a central role. Moreover, the growth of private equity and the tendency for start-ups to remain private for longer means that there are more private companies held by investors who need an exit, such as an IPO, at some stage.

Recent developments suggest change is finally on the horizon. The Securities and Exchange Commission (SEC) has authorized companies seeking to go public without raising additional funds to undertake a direct listing; a growing number of companies have gone public using an alternative structure in which a lead investor plays many of the roles long played by middlemen (although, thus far, often at a very high price); and better alternatives are on the horizon.[13] These alternative modes for taking a company public sometimes still use investment banks, but the investment banks involved function less like middlemen—exerting less control and earning lower fees. There has been some messiness in the process of transitioning away from having all companies follow the same well-worn path to having an array of options that can better accommodate the different needs of different companies. Nonetheless, that meaningful change has been so slow despite the significant cost savings and other advantages it could allow is yet another flag indicating excessive intermediary influence.

And, just as with real estate, the vectors allowing the traditional and expensive IPO process to survive as the dominant way for companies to go public for so long include repeat relationships among investment banks and between investment banks and sophisticated investors.[14] In theory, investors as a group have a lot to gain from liberating the IPO process from the grips of expensive investment banks, as they bear much of the burden of the high fees those banks charge. In practice, the repeat relationships between investment banks and sophisticated investors, as it plays out across multiple IPOs and other types of interactions, can make it hard for any individual investor to push back too aggressively. Among other things, they don't want to miss out on a chance to buy slightly underpriced and highly sought-after IPO shares.

Yet again, repeat relationships that can be helpful much of the time also have a downside. That the path for going public has remained too narrow for too long enabled investment banks to earn more at the expense of companies and investors. It may also have contributed to the decline in the proportion of large companies that are public, enabling even more secrecy and further reducing transparency and accountability.[15]

THE DARK SIDE OF INFORMATION AND RESOURCES

Another way that middlemen entrench and expand their dominance is through growth. Sometimes this is organic, as reflected in Walmart's systematic expansion across the United States and abroad. By bringing a wealth of information about where to place and how to design new warehouses and stores, supplying the resources needed to execute on those plans, and leveraging established relationships with suppliers, large middlemen can incrementally grow their footprint in ways that squeeze out previously viable alternatives, including local businesses and smaller chains. Acquisitions are another way middle-men expand their reach and power. For example, banks gobbling up other banks was a core mechanism through which the U.S. banking system evolved from one with more than 18,000 banks focused on serving their local communities to a system with fewer than 5,000 banks, in which just six banks hold the majority of bank assets.[16]

More inherently troubling are the ways that already power-ful middlemen use acquisitions to ward off or suppress threats that have the potential to displace them from their thrones. Amazon, once again, is a great example. In the 2000s, Soap.com and Diapers.com, both owned by Quidsi, rapidly grew in users and popularity. Internal Amazon emails reveal that executives there saw Quidsi as Amazon's biggest competitor in some domains. Amazon felt compelled to lower prices and offer better customer service as a result of consumers hav-ing an alternative that so many liked.[17] According to one assessment, Amazon was ready to lose $200 million a month selling diapers to prevent customers from shifting their buying to Diapers.com—a strat-egy akin to that used by those grocery stores in Chicago, but on a far greater scale and entailing far greater losses. Rather than continuing to win by offering better prices and service, Amazon soon took a differ-ent approach: In 2010, it decided it to buy Quidsi, thereby converting a competitor into part of its ever-growing empire.

Just the year before, Amazon had acquired Zappos, a popular online shoe store with a deeply loyal customer base. That acquisi-tion not only allowed it to take control of a site beloved by lovers of shoes, it also enabled Amazon to limit suppliers' ability to get access to those shoe-loving consumers without going through Amazon. For example, prior to Amazon's acquisition of Zappos, Nike had refused to enter into a distribution relationship with Amazon. Nike was just one of a number of Amazon holdouts that sold a lot of shoes through Zappos. In buying Zappos and eliminating a previously independent, alternative way for Nike to reach consumers who shop online, Am-azon put Nike in a position of feeling like it had little choice but to work with Amazon.[18]

Yet, focusing just on the competitive threat that Quidsi and Zappos posed at the time Amazon acquired them understates the impact of those acquisitions. Part of what has enabled Amazon to be so successful is that it is always looking ahead. It uses its remarkable wealth of data about what people are doing right now to plan what it should be doing next—and to understand possible threats to its dominance in the future. As antitrust experts Scott Hemphill and Tim Wu explain, acquisitions of nascent competitive threats—companies that may or may not be competitors today, but that could pose a serious threat in the future—can have a particularly pernicious impact on competition and innovation.[19] Thanks to Amazon buying Quidsi and Zappos when it did, no one will ever know just how great a threat to Amazon's dominance they may otherwise have become.

Amazon has also used acquisitions to disadvantage competitors in other ways. Recall Kiva, the robotics firm that Amazon acquired to increase the efficiency of its warehouses. Before being purchased by Amazon, the company sold robots to a number of e-commerce companies. And in the process of negotiating the merger, an Amazon executive assured Kiva's founder that Kiva could continue to sell robots to other companies, including Amazon's competitors. The founder had spent years cultivating those relationships and those customers had come to rely on Kiva, so this assurance was important to him. Yet, within a couple years of that acquisition, Amazon changed its mind (or revealed that it had contrary intentions all along), and it incrementally precluded Kiva from selling its robots to other companies, even long-term clients that relied on them.[20] This maximized the competitive advantage Amazon gained from the acquisition, but only because Amazon handicapped competitors alongside improving its internal operations.

Antitrust laws give the government the right to review and potentially block acquisitions that reduce competition. Yet, as a growing number of policy makers and scholars have shown, these laws have failed to fulfill their purpose in recent decades. One challenge is that regulators became mesmerized with economic models that purported to provide empirical assessments of the competitive impact of a transaction or action while too often ignoring the broader set of policy issues at stake and the limitations inherent in efforts to quantify power in rapidly evolving settings.[21] Another challenge is that whenever regulators do block or seek to impose meaningful conditions on a merger, the companies can and often do fight back in court. The courtroom thus becomes yet another setting in which large middlemen can use their deep pockets and deep understanding of the market in which they operate to weave tales about why a transaction will benefit consumers, even when their aims may be far less benign. That companies can readily fight back, while the consumers, workers, and others who may be harmed by a merger cannot, may accentuate the tendency of regulators to allow too many mergers to proceed.

As discussed further in the last chapter, recognizing that mergers have been used to fundamentally transform intermediation systems, such as banking, and to enable middlemen seeking dominance to further that aim, as reflected in Amazon, provides additional evidence of the value of rethinking and reinvigorating competition policy. The analysis throughout this book reveals that the power middlemen possess is both multidimensional—involving huge information advantages, relationships, and control over infrastructure and other resources—and dynamic, as middlemen use their current advantages to shape tomorrow's market structure. No one else could understand as well as Amazon the value *to Amazon* of acquiring Quidsi and Zappos. Similarly, the transformation of banking highlights the way examining mergers on a case-by-case basis can put regulators in a position of missing the forest for the trees. Banking and finance were transformed as a result of banks continually gobbling up other banks. Yet the current process for merger reviews leaves regulators narrowly focused on the competitive impact of each individual merger rather than the broader landscape.

The limited benefits that flowed from the multiple antitrust suits filed against the NAR and affiliates reveal how, even when antitrust authorities have good intentions and have found the right target, they still have not figured out how to design the type of remedies that can effectively address the distinct conglomeration of power that middlemen so often possess. In addition to revealing shortcomings in recent approaches to antitrust enforcement, these dynamics help to explain why the force that market fundamentalists see as key to ensuring companies are making society better off rather than just enriching themselves—competition—is so often failing to achieve that aim when the players in question are middlemen.

SHAPING THE LAW: THE DARK SIDE OF ORGANIZATION

If competition cannot keep middlemen in check, and enforcement of existing laws has not sufficed, the solution to the middleman economy might seem to lie in new laws that restore a healthier balance of power. Yet in the tug-of-war between middlemen and the little guy, lawmakers have not always sided with the little guy. Instead, lawmaking is yet another domain where middlemen are very good at harnessing their advantages to warp the process to serve their own ends.

Real estate again provides a vivid example. Trelora's business model is built on the premise that homebuyers should be able to choose whether they want an expensive, full-service middleman or whether they want to pay a fixed fee for a smaller bundle of services. The buy side is the area where Americans most overpay relative to peers abroad—in many places, buyers don't even use agents—making it a particularly important domain in which to give consumers access to a low-cost option like Trelora. Nonetheless, eighteen different states have at some point prohibited the rebates needed to enable competition on the buy side. Although a number of states have since repealed these laws, ten still have such laws on their books.[22] The reason: lobbying.

Lobbying describes efforts—whether by a company, an industry group like the NAR, or a public interest group, such as the Sierra Club—to influence lawmaking. Recent research shows that interest groups spend far more on lobbying than on campaign contributions, and that roughly 85 percent of the moneys spent on lobbying comes from corporations or trade group associations, like the NAR.[23] A variety of studies also show that lobbying can be effective, shaping tax policy, subsidies, and industry-specific regulations.[24] The anti-rebate laws that make Trelora's original business model unlawful are just one example of how these efforts can pay off.

In *The Logic of Collective Action: Public Goods and the Theory of Groups*, Mancur Olson laid the foundation for understanding why some groups are so effective at shaping policy, even when their interests run contrary to the public at large. His focus was on the challenges that impede effective collective action. In real estate, for example, there are far more homeowners than real estate agents, and those homeowners are better off if the law allows rebates. But most people buy homes only rarely, so they may not pay much attention to the small cost they incur from a law that prohibits rebates. Accentuating the challenge, homeowners have no readily available mechanisms to work together to promote their collective interest. And, even if they did, individual homeowners may be rationally apathetic, hoping to "free ride" on the work of others. As a result of these dynamics, there is often too little advocacy on behalf of laws that benefit broad, diffuse groups.

The situation is quite different for real estate agents. Anti-rebate laws help entrench a high-fee, two-agent structure so they have a much stronger incentive to fight for such laws. They are also better positioned to band together. As Olson explained, organizations that provide individualized benefits to attract members can then be harnessed to promote the common aims of those members. The NAR and local MLS fit this model perfectly. Real estate agents enjoy professional benefits from membership, but then become part of a lobbying behemoth. The NAR is the largest trade association in the country, with members in every single voting district, for state and federal purposes.[25] It has also been among the top five biggest spenders on lobbying every year since 2012, spending more than $40 million each year, and sometimes far more.[26]

To be a good lobbyist, however, requires more than just money. Information is key. Information in this context is not a neutral transmission of facts, but rather the selective presentation of certain facts to further a particular set of policies. Information can be used to highlight the advantages and disguise the drawbacks of a regulatory scheme that benefits middlemen. Information can also be used to sow confusion and uncertainty about what is best for the public. Because lawmakers are not experts in everything, those who are experts can shape how lawmakers see an issue. All of the informational advantages that middlemen use to help buyers and sellers connect can be redeployed to also shape policy. This is yet another flip side.

The proof is in the pudding. Not only have real estate agents convinced states to outlaw rebates, they have also succeeded in shaping federal law. In 2017, the NAR president bragged about how the NAR has long lobbied successfully to protect the mortgage deduction,[27] even though the deduction disproportionately benefits the wealthy.[28] He boasted about the way, in 2008, the "NAR successfully convinced Congress to shelve a planned tax increase on real estate partnerships"—protecting a deduction widely used by private equity and hedge fund managers to avoid paying income tax on what is effectively their income—because real estate agents also benefit from it. And when regulators proposed allowing banks or their affiliates to offer real estate brokerage services—potentially bringing meaningful competition that could benefit consumers—the NAR "worked hard to block the proposed rule," and got Congress on its side.[29] The overall results are striking. As NYU economist Lawrence White explains, "the United States' current housing policy is easily summarized in a few words: housing is good; more is better. And the way to get more is to subsidize it." Few industries benefit more from this support for housing than realtors.[30]

Lobbying by real estate agents even helps to explain FDR's decision to embrace home ownership as a top federal priority that merited massive subsidies. The NAR had been advocating such a policy for years.[31] The NAR understood that there was no better way to promote the collective interests of real estate agents than to increase the number of homes being bought and sold, and federal subsidies could do just that. Reflecting briefly on the alternatives that FDR might have embraced—such as making health care a right for every American or providing subsidies and infrastructure for child care—reveals just how much was at stake in that decision.

ORGANIZED, INFORMED, AND WELL POSITIONED

Once again, real estate is merely one embodiment of a trend that permeates intermediation more generally. According to Olson, there are two types of winners that regularly emerge in the collective action game. Some, like the NAR, are trade groups that provide individual benefits to members and lobby on members' collective behalf; the others engage in lobbying because they are so big that the individual benefits that they would enjoy justify investing in lobbying. Middlemen can be both types, and sometimes are both at the same time.

Trade groups are common. Groups such as the Institute of International Finance and the Bank Policy Institute conduct research and events that are useful to banks while also working hard to push laws that benefit banks. And banks such as JP Morgan Chase and Citigroup are so big that even on their own, they have a strong individual incentive to try to shape laws that affect their operations. According to OpenSecrets: "The financial sector is far and away the largest source of campaign contributions to federal candidates and parties."[32]

Retail middlemen are also big players. Between 1998 and 2008, Walmart went from spending a mere $140,000 a year on lobbying to more than $6 million a year on lobbying, and it has never looked back. Amazon came to the party later but has been ramping up its lobbying efforts in recent years, spending almost $17 million on lobbying activities in 2019 alone—more than any other single company in the country.[33] It also hasn't hesitated to hire the big guns, such as

President Barack Obama's former press secretary Jay Carney to work on its behalf.[34]

Again, however, it is not just the large sums that these middlemen spend to shape the law, but their ability to use that money in connection with their deep understanding of the markets in which they operate that make them so disturbingly influential.[35]

The evolution of the New York Stock Exchange in the 1970s provides another example of these dynamics in action. Initially, the rise of the NYSE as a centralized forum for buying and selling shares of companies listed on the exchange was a positive development. By bringing together buyers and sellers, it improved price accuracy and liquidity, which benefited investors and companies. But once the NYSE established itself as central, it became far harder for anyone to trade NYSE-listed stocks elsewhere, even if some NYSE rules benefited NYSE members at the expense of investors.

Just as with the real estate market, wealth transfers from investors to middlemen occurred via an artificially rigid pricing structure.[36] The form was a fixed-fee structure that required all NYSE members to charge the same price for a given trade. The similarities don't end there: Access and organization were key in both settings. Only NYSE members could trade on the exchange; this prevented price competition and transformed the NYSE into a powerful force lobbying on behalf of NYSE members. When the SEC pushed for change (thanks to prodding from the Justice Department), the NYSE fought back, utilizing both its organizational capacity and its informational advantages. The NYSE and its members claimed that competition could prove "destructive," leading to excessive consolidation that would harm small securities firms and some investors.

This put the SEC in a difficult spot: It had to choose between an arrangement known to work well and an alternative that could be better for investors, but that also entailed uncertainty. Eventually in 1975, after much delay and contestation, the SEC adopted rules that enabled securities firms to compete on price. The industry was transformed.[37] The rule change enabled the rise of discount firms that offered fewer services but also far lower prices and led to a more efficient and competitive market in which investors, on average, paid less to trade stocks. There was some fallout, and some firms did fail, but none of the horror stories spun by the industry came to fruition.

The delay between when such a competitive regime was viable and when the rule change finally went into effect is another testament to how middlemen use their control over collective infrastructure, organizational capacity, and superior information to entrench arrangements that allow them to earn overly high fees at the expense of end users.

Understanding how the NYSE, NAR, and their members have used their resources and market position to promote a pro-middleman regulatory agenda provides important insights into how the threat posed by large middlemen could grow in the years ahead. Although the influence of corporate lobbying may seem familiar, it takes on additional dimensions when the players involved are middlemen. Middlemen are not just experts in what they do, they also understand the needs of those they are trying to connect, such as consumers and retail investors. This enhances their ability to concoct stories about how a particular intervention could harm the very consumers or producers a law seeks to protect, and how consumers would benefit from legal changes that would really most help the middlemen involved. This makes it difficult for lawmakers to ignore their concerns, and makes it even harder for lawmakers and regulators to know what is true.

INADEQUATE FINANCIAL REGULATION AND OVERSIGHT

Inadequate regulatory regimes—that reduce the costs of doing business for financial middlemen while posing risks to ordinary borrowers and investors and the financial system more broadly—are further evidence of how successful financial middlemen have been in their lobbying efforts. For example, predatory lending was a major challenge in the early 2000s. The demand for loans that could be securitized, and the particularly strong demand for subprime loans, meant mortgage brokers had a strong incentive to foist such loans onto borrowers, even when they lacked the income to pay off the loan or when they could qualify for a traditional loan. Recall, Black borrowers were disproportionately targeted in these schemes, resulting in far more Black borrowers overpaying for mortgages ill-suited to their needs.[38] This prompted North Carolina, Georgia, and other states to adopt more aggressive consumer protection laws.

Banks found an ingenious way to push back. Most large banks are national banks, which means they are regulated by the Office of the Comptroller of the Currency (OCC), a federal regulator.[39] (Most smaller community banks, by contrast, continue to be state banks, overseen by state regulators and the FDIC.) As a federal regulator, the OCC enjoys a special power—it can exempt national banks from having to comply with certain state laws, so long as there is another relevant federal law that applies.[40] This type of preemption reduces compliance costs for banks, enabling them to have uniform policies across their operations. And, so long as the state and federal laws are equally tough, even if different in the particulars, not much is lost.

But in the early 2000s, when some states had enacted more robust consumer protection laws while Congress and federal regulators had not, there was a meaningful difference in the degree of protection afforded by the state and federal regimes. Nonetheless, in 2003, the OCC proposed a regulation that would shield national banks from the new state laws and other state-law-based consumer protections. Consumer advocates opposed the rule, contending that it "would expose consumers to wide-spread predatory and abusive practices by national banks."[41] In response, national banks and their trade groups devised a new narrative. They argued that preemption would benefit consumers. According to the banks, the complications and uncertainty banks would face if forced to comply with a patchwork of state laws would deter banks from lending in certain markets and providing loans to riskier borrowers. They used their information and resources to spin a potentially plausible tale that the state laws meant to protect consumers were the by-product of naiveté and a failure to understand the benefits of innovations such as securitization and the importance of standardization in facilitating securitization. It was a classic consumer-versus-middleman battle and, as so often happens, the middlemen pretended that what was best for them also benefited consumers, and the government sided with the middlemen.

With the benefit of hindsight, the OCC's reasoning could not look more flawed. As law professors Patricia McCoy and Kathleen Engel demonstrate, the OCC's approach meant many borrowers "had virtually no remedy against abusive lenders."[42] Making matters worse, the increases in home ownership proved short-lived. Many of those who bought their first home in the 2000s were forced out of those homes as a result of the financial crisis. The boom and bust was particularly pronounced for Black households, who lost almost all of the ground that they had appeared to gain in the run-up to the crisis. A Harvard study found that the disparity between Black home ownership rates and white home ownership rates was even bigger in 2017 than it had been back in the mid-1990s.[43]

This was typical of the increasing power held by large, powerful financial institutions in the decades leading up to 2008, which gave those middlemen outsized influence in legislative and regulatory processes. This contributed to insufficient protections for consumers, deregulation, and inadequate reforms even as the financial system evolved in ways that posed new threats.

This comes through not only in isolated examples but also through broad trends observable over time. For example, starting in the 1980s, the Fed and the OCC incrementally chipped away at the Glass-Steagall wall separating banks and investment banks. Prodded by the banks, these regulators started allowing banks to engage in an increasing array of risky, investment-banking-like activities, and they typically did so with little regard to the way such changes might alter the competitive playing field or what other changes banks should have to undertake in light of the new risks to which they could now be exposed.[44] Similarly, asset managers, another type of middlemen, sought to create a new type of investment fund that would replicate deposits without anyone actually having to pay for deposit insurance—the now-pervasive money market mutual funds. The SEC paved the way for them to do so and no one else stopped them. This set the stage for the growth of a massive system of market-based intermediation, commonly known as the shadow banking system, that undertook many of the same functions as banks but was not regulated to address the systemic threats that it posed.

Congress, a major recipient of campaign contributions and a regular target for lobbying efforts, also played a role. When the Commodity Futures Trading Commission (CFTC) suggested it might regulate derivatives, for example, Congress passed a law effectively exempting derivatives from regulation.[45] When banks sought to further break down the Glass-Steagall divide between banks and investment banks, Congress went along, facilitating the growth of overly large, complex, too-big-to-fail financial institutions.[46] Congress also contributed to the growth of short-term funding markets that are a frequent source of financial fragility by exempting short-term sale and repurchase agreements (commonly known as repo transactions) from core bankruptcy protections.[47] Without these exemptions, the repo market would be far smaller. And as we will see in the next chapter, the decisions by Congress and regulators to side with financial intermediaries, time and again, ultimately proved disastrous for the entire country and much of the world. The rules governing banks and other financial intermediaries were strengthened significantly after the financial crisis, and the Dodd-Frank Wall Street Reform and Consumer Protection Act of 2010 addressed many of the specific abuses described here, but the very fact it took such an unmitigated disaster to prompt lawmakers to adopt adequate regulation is further evidence of how skewed the balance is most of the time.

CHAPTER 8

THE MYTH OF SUPPLY CHAIN ACCOUNTABILITY

O N SEPTEMBER 18, 2007, a sunny fall day in Washington, D.C., Federal Reserve chair Ben Bernanke gathered with his colleagues on the Federal Open Market Committee, better known as the FOMC. Responsible for the nation's monetary policy, the FOMC wields an enormous amount of power. The meeting had taken on new importance as a result of problems in the market for subprime mortgages. The real source of concern, however, was the way problems in that small market seemed to be spilling over and triggering dysfunction in other parts of the financial system.

The FOMC had already taken the unusual step of reducing interest rates between regularly scheduled meetings, but this was the first time they had gathered in person to assess what had gone wrong and what might lie ahead. The conversation quickly revealed that securitization had played a central role in contributing to the problems, and that it was also impeding the ability of Bernanke and his colleagues to see how risks were allocated across the financial system and what the future might hold.

As Federal Reserve governor Randall Kroszner explained: "With the originate-to-distribute model and securitization, . . . risks are much more dispersed."[1] This creates "pockets of uncertainty, and that is exactly what has come up. People don't have as much information as they thought they had."[2]

Supposedly sophisticated investors were among those that "[didn't] have as much information as they thought they had," contributing to the dysfunction. Recall, one "benefit" of securitization is that it divvies up both risks and information burdens, enabling the creation of some MBSs that were so low risk that they were supposed to be "information insensitive." Rating agencies aided this transformation, providing investors the additional assurance that a large share of the MBSs and other securitized assets issued had sufficiently little risk as to merit a AAA credit rating. Unfortunately, many of those ratings proved to be far from accurate.

Over the summer of 2007, the major rating agencies downgraded numerous MBSs backed by subprime and other atypical mortgages. But because of the complexity of the supply chains through which money flowed from mortgages to MBSs, it was difficult for investors to readily ascertain what these instruments were worth if they couldn't trust the ratings. This resulted in a lot of investors eager to sell subprime MBSs but very few who were willing to buy them. As the French bank BNP Paribas explained in August when it suspended redemptions in three funds exposed to MBSs: "The complete evaporation of liquidity in certain market segments of the U.S. securitization market has made it impossible to value certain assets fairly, regardless of their quality or credit rating."[3] With no one buying, there were no trades, and therefore no easy way to price the MBS assets held by the funds.

What happened next caught everyone off guard. The announcement triggered worries about the actual value of all kinds of securitized products. Having lost faith in ratings and lacking the tools and information needed to assess how higher default rates among borrowers would affect the various tranches of MBSs issued and the array of other instruments backed by or exposed to those MBSs, the intertwined problems of too few buyers, too little liquidity, and market dysfunction spread.

William Dudley, who at the time oversaw implementation of the Fed's monetary policy, provided the FOMC a report on the range of unanticipated ripple effects emanating from the problems in the subprime MBS market.[4] For one, banks and other mortgage originators were having a hard time securitizing any mortgage that did not have a federal government backing. Because many banks had come to rely on securitization to move new loans off their balance sheets, the sudden freeze in the securitization market was creating strains on their balance sheets, limiting their ability to make any new loans, even to the most creditworthy borrowers.

Dudley further explained that there were also significant disruptions in short-term funding markets. Another newfangled financial innovation—asset-backed commercial paper (ABCP) conduits—was a primary culprit. These were a prime example of the complicated capital supply chains of the era. In simplified terms, a bank would create a new, off-balance-sheet entity, such as a special purpose vehicle, that it would use to buy up MBSs, which it would fund by issuing short-term commercial paper backed by those assets—the ABCP— and some longer-term instruments that could absorb losses.[5] Money market mutual funds were the biggest buyers of that ABCP, and for a long time they just kept rolling over their holdings. The overall system functioned much like a bank, with money market mutual fund shares playing the role of short-term deposits that fund long-term home loans, but with a lot more layers and complications. This was the heart of shadow banking circa 2007.

The extra complications mattered little when times were good but a lot when conditions soured. Once questions arose regarding the value of subprime MBSs and the reliability of credit ratings, money market funds—that promise safety to their investors—wanted nothing to do with ABCP backed by such assets. According to a report by the New York Fed, 40 percent of ABCP programs experienced a "run" by the end of 2007.[6] This caused the trouble to spread to the banks that had sponsored those programs.

Just as significant as the gaps in what investors knew were the gaps in what regulators knew.[7] Precisely because of the inherent fragility of any institution that relies on short-term liabilities, such as deposits or commercial paper, to fund longer-term assets, such as loans, banks in the United States have long been subject to significant regulation and oversight. There is no comparable regulatory regime for the shadow banking system, hence the term "shadow." This was the point that Governor Kroszner was making to his colleagues: With the rise of securitization, neither investors nor regulators knew precisely how big the risks were or how they were dispersed.

The long, complex securitization chains that were the backbone of this new system of market-based intermediation—the shadow banks—appeared to create efficiencies when they were proliferating in the early 2000s. They seemingly made it easier for people to realize their dream of home ownership by enabling those seeking a home loan to tap into new pools of capital. The system appeared to be yet another manifestation of the ways that hyper-specialization enabled by long and complex supply chains could make people better off, just as with clothing and food production.

But in ways no one appreciated until the whole system blew up, taking the economy along with it, the effort to create short-term efficiencies through layers upon layers of middlemen and ever more complex capital supply chains also created an excessively rigid, fragile, and opaque system. Like many complex supply chains, short-term gains had come at the expense of simplicity and the resilience and accountability it enables.

The results are all too well known. The gyrations that so concerned the Fed in the fall of 2007 proved to be mere tremors relative to the earthquake to come. The failure of the investment bank Lehman Brothers and the near failure of insurance giant AIG—which managed to avoid bankruptcy only because of an $85 billion emergency loan from the Fed—rattled financial markets the world over.

The financial crisis that ensued inflicted deep wounds on working Americans and the U.S. economy. More than nine million families lost their homes through a foreclosure or distressed sale.[8] Unemployment reached 10 percent, not including the many Americans who gave up on trying to find a job.[9] Housing values fell by 30 percent and the major stock market indices fell by half. Everyone saw their wealth shrink.

Making matters worse, while the wealth of families in the top 10 percent recovered quickly in the years following the crisis, the less affluent continued to suffer.[10] Black families experienced even greater declines in their home equity, home ownership rates, and wealth than white families.[11] The crisis also inflicted lasting scars on society and exacerbated political divisions. The common perception that the government had saved Wall Street while allowing Main Street to suffer created widespread outrage and helped feed a growing populist movement.

All of this happened even though Fed officials and other policy makers had a full year between the September 2007 FOMC meeting when it learned of the core problems and the September 2008 failures of Lehman and AIG. It was in 2007 that Janet Yellen, then president of the San Francisco Fed, warned her Fed colleagues: "the most important factor shaping the forecast for the fourth quarter and beyond is the earthquake that began roiling financial markets in mid-July." It was in 2007 that Federal Reserve governor Frederic Mishkin warned of "the potential for a vicious circle or a downward spiral" in which "a financial disruption" makes "it harder to allocate capital to people with productive investment opportunities," leading to "a contraction in economic activity."[12] That policy makers remained so unprepared a full year later is not a sign that they disregarded the threats; it is a testament to how difficult it was for anyone to figure out the many different interconnections through which problems could spread because the system had gotten so complicated.

The year that passed between when the dysfunction first set in and when Lehman failed, and the forgone opportunity to avoid the subsequent catastrophe, is a testament to how hard it is to shift from a system based on proxies, such as credit ratings, to one based on *actual* information once intermediation chains become too long and too complex. The complexity of securitization chains not only added to the tendency of investors to flee, instigating fragility, but it also reduced the capacity of the Fed and other regulators to fill in the gaps in what market participants knew, exacerbating the fragility.[13]

Although different in the details, the dynamics are akin to those at play in Germany's 2011 *E. coli* outbreak. In both instances, there were early signs of trouble that posed a serious threat to public well-being. But the officials charged with protecting the public faced such a complex morass that they could not readily discern the full nature or origins of the threat they were facing.

In each instance, these information gaps contributed to public officials making mistaken diagnoses with adverse consequences: German public health officials prematurely and inaccurately claimed that Spanish tomatoes were causing an outbreak that was in fact caused by sprouts grown in Germany. Similarly, in 2008, Fed officials demurred an offer of the very regulatory authority that may have allowed them to avert the failure of Lehman Brothers and the wreckage that followed. After Bear Stearns failed—and more than five months before Lehman did so—Senator Chris Dodd asked Bernanke whether the Fed needed greater authority over broker dealers such as Bear Stearns and Lehman Brothers. Bernanke said no.[14] This was but one of numerous missed opportunities, arising less out of bad faith than a fundamental misunderstanding enabled by a lack of information. Just like the *E. coli* outbreak in Germany, the magnitude of the problem grew in size and scope, inflicting far more harm on ordinary people as a result of blind spots that arise as supply chains grow too complex.

COMPLEX SUPPLY CHAINS, FRAGILE AND RIGID

Alongside creating debilitating informational challenges, the excessive complexity of the capital supply chains that dominated in the mid-2000s also created harmful rigidities. As housing prices plum-meted, many homeowners found that they owed more on their mort-gage than what their house was worth. At the peak, in 2010, fifteen million American families found themselves "underwater."[15] This not only caused a huge amount of distress for those homeowners but also created challenges for the investors exposed to those loans.

When the supply chain was relatively simple (borrower–bank–depositor), the borrower could go directly to the bank and the bank could and often did renegotiate. Because homeowners always have the ability to walk away if loan terms are too onerous relative to the value of a house, reducing the amount owed can sometimes increase the ex-pected performance and value of a loan. But when a borrower instead had to go to a servicer and that servicer had to consider the interests of numerous different tranches of MBSs—which would be affected differently depending on which terms in a mortgage were changed—it became exceptionally difficult to modify mortgages in the ways that would have been most helpful to borrowers and therefore most effec-tive at averting foreclosures.[16] The net result was too many foreclo-sures. This was bad for homeowners, bad for MBS investors, bad for the neighborhoods where those homes sat, and bad for the economy.[17] Once again, the supposed efficiencies when times were good created excessive rigidities that hurt those involved, and others, when housing markets declined.

Another short-lived but vivid example of the impact of supply chain rigidities arose in March 2020. When it became clear that Covid was in the United States and spreading rapidly, lifestyles changed quickly and all in the same way. People who normally would use the bathroom at their places of employment or school five days a week now faced the prospect of going only at home. Anticipating the change and wanting to avoid too many trips to the store, Americans stocked up on toilet paper, along with rice and frozen veggies.

But retailers that had spent years reducing inventory to enhance ef-ficiency couldn't keep up with the increased demand. Making matters worse, uncertainty about when more toilet paper would be available led to a panic, resulting in even more buying, and even greater short-ages. By March 23, 2020, 70 percent of U.S. grocers (including online outlets) were completely out of stock.[18] The quest for toilet paper be-came a national pastime. Former U.S. congressman Brad Miller took to Twitter to declare his day a success when he managed to "score 20 rolls of toilet paper" at his local Safeway. Buried beneath such banter was a recognition of common vulnerability. As if the shock to pub-lic health and the economy were not destabilizing enough, access to daily necessities could not be assured. For many, the struggle to find a roll of toilet paper exacerbated and embodied the emotional distress of the moment.

Yet the toilet paper shortage of 2020 proved to be a mild precursor of the supply chain challenges to come. As the economy roared back to life in 2021, so did demand. People wanted to build and renovate houses, increasing demand for lumber. But the sourcing and inter-mediation systems for lumber couldn't keep up. The price of lumber skyrocketed 323 percent between April 2020 and May 2021, leading to delays and frustration. The price of lumber soon came back down, but the volatility merely added to the uncertainty, making it difficult for builders and families to plan.

A shortage of semiconductors—"[t]he 'brain' within every elec-tronic device in the world"—also reached crisis proportions in the spring of 2021.[19] Apple had to delay the introduction of its newest iPhone; Ford and other auto companies anticipated forgone prof-its in excess of $2 billion each, stemming from the shortage; and even the U.S. Department of Defense grappled with how to obtain this critical input. There were also ripple effects on other markets. The reduced production of new cars, for example, contributed to increased demand for used cars. In June 2021, a widely used index of used car prices was up 36 percent from the previous year, and some used cars were selling for more than the price they had fetched when they were new.[20]

The length of today's supply chains was not the only factor con-tributing to these shortages and price fluctuations, but it played a major role. The multi-nodal, multi-continent, multi-company mode of production created dependencies, vulnerabilities, and massive information gaps. The companies that most people think of as manufacturers and makers were revealed to be merely the last notch in a long supply chain over which they had limited control and visibility. The huge delays and losses were reflective of the way that companies from Apple to Ford had failed to understand the upstream vulnerabilities that arise from such disaggregation. Exacerbating the challenge, taking a single process and disaggregating it across nodes left each node to fend for itself. This led to decisions that seemed optimal at an atomistic level but created adverse systemic ripple effects, such as underproducing semiconductors until companies further up the chain committed to buying them. And by that time, production could be ramped up only so much. Interdependencies across supply chains further accentuated the fragilities and rigidities. For example, lumber producers had a hard time ramping up supply because of supply chain issues in the manufacturing equipment they required.[21] The dynamics resembled the challenges that arose around securitization in 2008: Efforts to maximize short-term "efficiencies" had resulted in separations that reduced costs when everything went smoothly but magnified costs, uncertainty, and dysfunction once a shock hit.

Global shipping during the recovery accentuated and exemplified these challenges. As economist and historian Marc Levinson shows in his book *Outside the Box: How Globalization Changed from Moving Stuff to Spreading Ideas*, as making became disaggregated, "international businesses were slow to grasp the ways in which their new business model had created new risk," and one of those risks was being far more vulnerable to changes in the cost and availability of transit.[22] By the spring of 2021, shipping costs had increased 25 to 50 percent, and more in some cases, relative to pre-pandemic rates.[23] Companies also faced massive delays in actually getting goods where they needed to be. In March 2021, only 40 percent of all container ships arrived in port on time.[24] This made it far more difficult for manufacturers and middlemen to know what they could expect to receive and when. As one expert opined, this motivated many to over-order relative to their actual needs in hopes of just getting some of the inputs and products they required. Yet again, this may have been the optimal strategy for these actors, but it also increased the probability of yet further price dislocations and further shifts from undersupply to oversupply and back again.[25]

Just as in 2008, these dynamics adversely affected the lives of ordinary Americans, who faced rapidly increasing prices and a complete inability to get certain goods. They also created a host of challenges for policy makers. The lumber shortage revealed the extent to which U.S. home builders and buyers were exposed to Canadian logging policies. The semiconductor shortage set off alarm bells as foreign policy experts realized that the ten biggest manufacturers were all in Asia, including two in Taiwan and two in China. That tensions with China were already simmering—in part because of different views regarding the autonomy of Taiwan—created a risk that U.S. leaders might have to make difficult trade-offs between the economic health of U.S. companies and consumers, and pursuing an agenda driven by ethical and other policy concerns.

The way supply chain weaknesses reverberated through the economy and affected prices also created challenges for the Federal Reserve. Long charged with promoting the dual aims of price stability and full employment, the Fed had revamped its monetary policy framework in 2020 to allow the economy to "run hot" after periods when inflation was persistently below the Fed's 2 percent target, as it had for much of the decade before the pandemic. Many hoped this new framework would allow the Fed to maintain a more accommodative monetary stance coming out of the pandemic, resulting in more jobs and higher wages for more workers. The supply chain disruptions and the price hikes they wrought cast a dark cloud over this plan. In May 2021, for example, the Consumer Price Index was 5 percent higher than the previous year, the biggest jump in nearly two decades.[26] The bigger challenge was that Fed officials had not expected the jump, and could not readily determine the cause. If it was just supply chain challenges that would rectify themselves, they could forge ahead with their plans and help get more people back to work. But if those price increases were reflective of more fundamental shifts, then delaying tightening, some feared, might lead to too much inflation.[27]

Although the specifics vary, the overarching trend is the same: Aspects of the middleman economy that seemed to create short-term efficiency gains simultaneously created latent vulnerabilities. When those vulnerabilities gave rise to dysfunction and uncertainty, ordinary Americans suffered and policy makers were hampered in their ability to know how best to help.

SUPPLY CHAINS AND ACCOUNTABILITY

One reason the 2008 financial crisis set off a political revolution was the sense of unfairness it triggered. It wasn't just that big banks got rescued while homeowners suffered. It was that the executives in charge of those banks were not held accountable by prosecutors for the roles they had played in contributing to the fiasco.

The dearth of prosecutions cannot be attributed to a lack of bad behavior. Between 2007 and 2008, the FBI investigated more than 2,800 cases of mortgage fraud, nearly half of which involved losses in excess of $1 million.[28] At the other end of the supply chains into which these mortgages were fed, there were numerous accounts of investment banks selling MBSs and CDOs that they either knew or should have known were backed by low-quality mortgages.[29] One investigation revealed that Merrill Lynch told investors that the loans it was securitizing were made to borrowers who were likely and able to repay their debts even though its due diligence regularly revealed numerous underwriting and compliance defects.[30] This was among the many findings of bad behavior that ultimately resulted in a $16.5

billion settlement between Bank of America (which by that time owned Merrill Lynch) and the Justice Department. Other banks entered into similar settlements. By 2017, banks collectively had been forced to shell out more than $200 billion in fines and legal fees for the various ways they had enabled and profited from bad behavior in connection with originating and securitizing home loans.[31] Numbers this big suggest large-scale failures and pervasive wrongdoing. Nonetheless, not a single senior executive at Bank of America or any other major bank faced charges. Nor did any of the executives at Lehman, Bear Stearns, or AIG face criminal liability for running their firms to the ground.

The disconnect between the seemingly rampant fraud in mortgage origination practices and MBS sales, on the one hand, and the lack of criminal prosecutions of bank executives, on the other hand, is another example of how long, complex supply chains undermine accountability. Despite plentiful evidence of bad behavior by low-level employees, it was hard to show that high-level executives knew enough about what was happening to be held liable alongside the companies that they ran. This is far from the only explanation—as Jennifer Taub and Jesse Eisinger have shown in two wonderful books, there are meaningful, systemic shortcomings in the willingness of prosecutors to pursue white-collar criminals—but it helps to explain the mismatch between the bad behavior that occurred and the dearth of executive prosecutions.[32]

The lack of mechanisms beyond legal liability for holding executives accountable is itself indicative of the shortcoming of allowing such a massive gulf to separate those borrowing funds and those supplying them. Back when community banks and thrifts dominated the landscape, borrowers, depositors, and bank employees were often all part of the same community. This made it harder for anyone to lie and created additional incentives for honesty and good behavior. Although there has always been some chicanery in finance, the dark shadows cast by complex structures enable a whole lot more of it.

SUSTAINABLE JEANS

Since I have no aspiration of ever sitting in a corner office, it is easy for me to criticize the lack of executive accountability. But there are times that I too take shelter in the ignorance that long supply chains can foster. For example, I know that inexpensive clothes are often the by-product of cotton farmers and garment workers laboring away for too little pay in unsafe conditions, and often entail environmental harm. Yet, that only sometimes stops me from buying them.

In this regard, I am not alone. As we have learned, people today buy more clothes and shoes than they did even a decade ago, and far, far more than their grandparents. In addition to using fancy data analytics to encourage people to buy more, the middleman economy facilitates this extra consumption by shielding buyers from seeing the effects of these buying decisions.

One of the factors that behavioral economists have identified as critical to real-world decision making is salience.[33] This helps to explain why people eat cake even when they want to lose weight, and why people who care about the environment may nonetheless drive large, gas-guzzling vehicles and take long showers during a water shortage.[34] The immediate, tangible joy of the experience is more salient to them than the abstract, probabilistic impact of their actions on their health or the environment. For better and worse, most of us overvalue what we see and experience and discount what we don't.

One challenge with behavioral biases is that knowing about them doesn't free us from them. When I see a bar of chocolate at checkout, I am often still tempted to buy it despite what I know about cocoa labor practices. Similar issues arise when I shop for new T-shirts and jeans for my family. In the moment, I instinctively still prioritize price, convenience, style, and comfort. When that happens, I often compromise on other values that I would like to think are more important to me, such as respecting the dignity of all human beings and our planet.

The problems start with the cotton that goes into those T-shirts.

Picking cotton has always been laborious, which is one reason that the work is so often delegated to those who have the least amount of control over their circumstances. According to the U.S. Department of Labor, seven of the top ten cotton-producing countries use forced or child labor in their cotton industries.[35] The United States, Australia, and Mexico are the only top producers that don't; and, with the use of prison labor in parts of the United States, some wonder whether it too should be on the list.

The cotton then needs to be sorted, made into thread and then textiles, dyed, and transformed into T-shirts and other finished products. Questionable labor practices pervade each of these nodes.[36] For example, numerous reports suggest that China has been relying on forced labor by 1.8 million Uyghurs and other Muslim minorities in its cotton production and thread/yarn, textiles, and garment industries.[37] Companies identified by U.S. authorities as selling clothes tainted by Uyghur labor include H&M, Nike, and Patagonia.[38] These are brands many people trust, in part because at the time of the report, each company already had policies in place that purported to address labor issues along their supply chains. At the time the U.S. Department of Labor identified Nike as a seller of tainted goods, for example, Nike had a policy declaring: "we have a responsibility to conduct our business in an ethical way," and "[w]e expect the same from our suppliers." It also had a detailed "Code of Conduct" purportedly meant to prevent its suppliers from using forced labor or engaging in other wrongful behavior.[39] Nonetheless, those policies failed to prevent money flowing from buyers of Nike goods to Nike to suppliers that used forced labor.

Even when garment workers supposedly enjoy the autonomy denied to Uyghurs in China, they still often face hazardous working conditions. This was vividly illustrated in 2013, when the collapse of the Rana Plaza building in Bangladesh killed more than 1,100 people and injured another 2,500, mostly garment workers.[40] That Bangladeshi garment workers faced unsafe working conditions had already been well documented. More than five hundred garment workers had perished in factory fires in Bangladesh in the six years before the collapse.[41] Yet local laws did little to protect them. Nor did it require that the workers be compensated for the risks they were enduring. At the time, garment workers could be paid as little as $68 a month, a fraction of the $280 a month minimum wage that even China requires for most of its garment workers and an even smaller fraction of what those same people would need to be paid if lawfully employed to do the same work in the United States. These are the conditions that give me the option of buying such inexpensive T-shirts for my girls, but I don't see any sign of this when I peruse online or in the store.

Middlemen and long supply chains also help blind consumers to the environmental impact of their purchases. According to a report from the United Nations, the fashion industry is responsible for 8 to 10 percent of global carbon emissions and 20 percent of global waste water.[42] It takes 2,000 gallons of water—roughly the amount of water one person consumes over seven years—to make just one pair of jeans.[43] The production of cotton is also responsible for 24 percent of insecticides and 11 percent of pesticides used worldwide.[44] And because so much of the garment industry operates in parts of Asia that still rely on coal and natural gas, the industry is expected to account for a whopping 50 percent of greenhouse gas emissions by 2030, if it remains on its current trajectory.[45] Just as with labor, none of this comes through when I go shopping, and middlemen probably rack up more sales as a result.

One potential response to these challenges is to require companies to provide consumers more information so that the impact of their choices is not as hidden from view. Precisely because so much policy making has been informed by economics in recent decades, and disclosure requirements appear to give both consumers and companies a lot of autonomy to make decisions that suit their individual preferences, these types of interventions have been popular and widely adopted. If they worked as intended, such rules could reduce the need for structural reforms to tackle the problems revealed here. In practice, however, the evidence suggests disclosure is rarely as helpful as advocates claim in addressing policy challenges, and the middleman economy is no exception.[46]

WARLORDS AND IPHONES

One well-intentioned effort to use mandatory disclosure requirements to ferret out bad behavior and give American consumers and investors more information about the impact of their actions is the "conflict minerals rule." According to Republican senator Sam Brownback, one of the original sponsors of the bill introducing the rule, he and Democrat Russ Feingold found "[a] humanitarian crisis of incredible proportions" when they visited the Congo together.[47] He explained that millions of people in the Congo and neighboring countries had died in the past decade, that rape and other forms of violence were rampant, and that American consumers could be contributing to the challenges because the rebel groups perpetuating the violence funded their activities, in part, by mining and selling minerals that made their way into cell phones, laptops, and other goods imported into the United States.[48]

Congress sought to address the problem through up-the-supply-chain disclosure requirements. More specifically, Congress decided that all public companies that used minerals of the kinds sourced in Congolese mines (mainly gold, tantalum, tin, and tungsten) should provide greater information to investors about whether they were or might be buying minerals from mines that funded Congolese rebel groups.[49]

Efforts to implement the rule revealed that most companies knew very little about the origins of the minerals they were using. In 2014, only 33 percent of companies could even identify the countries from which they sourced minerals.[50] After implementation of the rule, far more companies could identify their minerals' country of origin, but they still had a hard time discerning whether they were supporting rebel mines.[51]

Given these difficulties, many companies subject to the rule opted to avoid buying any minerals from the Congo, whether mined by rebels or other locals. As a result, minerals coming out of the Congo often sold for far less than the same minerals sourced from other places, hurting many of the Congolese people the rule was meant to help.[52] According to one estimate, "1–2 million Congolese artisanal miners" were affected directly and "up to 5–12 million Congolese civilians" have experienced indirect harms as a result of the rule.[53]

The lawmakers who advocated for the rule had good intentions, and there have been some salutary effects, including a decline in the number of mines controlled by rebel groups. The law has also contributed to making companies more diligent in monitoring their supply chains. Apple, for example, which was often the face of the debate, has removed over a hundred smelters and refiners from its supply chain for refusing to be audited.[54] And complementary efforts by the U.S. Department of State and the U.S. Agency for International Development in the area, including efforts to certify legitimate mines, may produce long-term beneficial outcomes. But there were also a lot of unintended consequences. While only one example, it illustrates why disclosure mandates, on their own, rarely achieve the same level of accountability that comes from shorter supply chains.

FROM CREDIT RATINGS TO GREEN BONDS

Even without government mandates, many companies want to provide information about their labor practices, sustainability efforts, or quality as a way of attracting customers and investors. Given that such indicators are often used to justify a higher price tag, there are also a host of institutions that purport to set standards or help verify these types of claims. The idea is to rely on trustworthy, independent experts to bridge the information gaps that arise as supply chains and intermediation regimes become long and complex.

The events leading up to the 2008 crisis again serve as a useful starting point. Investors buying up MBSs were not blindly relying on the claims made by MBS sponsors; they were also relying on credit ratings. Credit rating agencies are meant to be neutral and trustworthy. In theory, because their ongoing viability depends on investors trusting the ratings they issue, they should have a good incentive to provide accurate assessments. They are also experts in assessing credit risk. This helps to explain why investors, and even regulators, so often rely on ratings.

This system worked reasonably well when the agencies were assessing the health of individual companies and rating the debt they issued. But the process of packaging diverse assets together and creating tranches of new financial instruments to more finely dice, slice, and allocate risks greatly complicates the task of assessing the probable performance of the debt issued. As MBSs got more complicated and additional layers of securitization vehicles were added by CDOs, the task became more complicated still. As the events of 2007 and 2008 revealed, credit rating agencies were not very good at this undertaking. There were a lot of reasons for their failures, including conflicts of interests and excessive reliance on limited historical data, but a core challenge was that the layered supply chains had just gotten too complicated.

The complexities of these structures have also undermined efforts to hold credit rating agencies more accountable and to reduce reliance on them. Lawmakers adopted a range of reforms after the 2008 crisis, but with only mixed success. Most investors and regulators just don't have the resources to assess how something as complex as an MBS is likely to perform without outside guidance. And, so long as investors are willing to pay more for assets deemed safe by a credit rating agency, issuers of debt will have an incentive to game the system to produce as much supposedly safe debt as possible.

Similar dynamics are playing out in a host of other domains for related reasons. First, investors and consumers are increasingly willing to pay a premium for assurances that they are supporting fair labor, sustainability, or other features not apparent from a product itself. Second, because it is hard for investors and consumers to assess these matters for themselves, they often rely on third parties that purport to verify an investment is green, food is organic, or clothing is produced in a sustainable way. And third, the ability to earn that premium often results in gamesmanship. Companies, for example, find ways to take just enough rightful actions to earn a favorable rating while still avoiding the more costly steps needed to live up to the spirit of what the rating scheme is meant to assess. To see how this happens, each step is considered in turn, first in the consumer arena and then in finance.

First, recent empirical evidence shows a growing proportion of consumers will pay a premium for consciously produced goods. For example, in 2014, researchers undertook a large-scale survey to assess whether Greek consumers—who regularly say in surveys that they care about issues like worker protections—were really ready to pay more for assurances that workers were treated appropriately.[55] They focused on strawberries, as a good commonly associated with inexpensive immigrant labor, and used techniques specifically designed by other researchers to elicit more accurate assessments of people's actual willingness to pay. They found that consumers, on average, were willing to pay 70 percent more for strawberries when assured that laborers were paid fairly for their work. Although premiums vary, a number of other studies have similarly found that consumers are willing to pay more for goods produced in a more ethical way.

Yet, second, apart from settings such as CSAs and farm stands that allow consumers to interact directly with farms and farmers, consumers have no ready way to assess whether laborers were in fact paid a fair wage or what chemicals may have been used to ward off insects and maximize yields. This is a drawback of the middleman economy: Consumers cannot see any of this for themselves.[56]

This leads to the third issue—whether third-party certification schemes can effectively bridge the gap. The evidence here is mixed. On the one hand, some studies find that third-party certification schemes can convey meaningful information and induce better labor and sustainability practices. For example, many Central American coffee farmers successfully used third-party certification schemes to maintain better practices even in the face of increased price competition from coffee plantations in places such as a Brazil and Vietnam, starting in the 1990s. To avoid having their beans treated as just another commodity, they differentiated themselves by offering a higher-quality and more sustainable product. Established third-party certification systems, namely Fair Trade and Rainforest Alliance, along with organic certification, were key to their success.

Subsequent research has confirmed that coffee growers could earn a real premium for certified organic coffee, and that in order to qualify as organic, farms made meaningful changes to their production processes.[57] Yet it is not clear whether these approaches meaningfully enhanced the quality of life enjoyed by farmers. The additional costs incurred to earn an organic certification significantly reduced any pecuniary gain to the farmer. Moreover, while Fairtrade coffee in particular could be sold at a much higher price, the demand was limited. This meant farmers often had to opt into the Fairtrade regime and incur the associated costs without any guarantee they would be able to sell their coffee beans at the higher prices Fairtrade was able to pay.

Other research casts even more doubt on the utility and informativeness of certification efforts. A core challenge is that many of the studies that suggest Fairtrade is beneficial are funded by the very companies seeking to use certification schemes to further their own business interests.[58] Wanting an independent assessment of these schemes, the UK Department for International Development funded a four-year research project to assess working conditions and poverty rates in Ethiopia and Uganda, and the impact of Fairtrade efforts on both.[59] With the government's financial support, academic researchers designed a rigorous study that used surveys, interviews, and other assessments of compensation and conditions, and they devised appropriate reference points, for example, comparing two large-scale flower growing regions in Ethiopia, one with and one without Fairtrade.

For anyone who has ever felt good about buying something

because of a Fairtrade certification, the results are disheartening. In three of the four domains studied, wage laborers were far more likely to be paid *less* than 60 percent of the median wage (the study's cutoff for a respectable, if very low rate of, pay) in the Fairtrade sites. The Fairtrade sites typically had greater pay differentials, with higher-paid workers earning more but lower-paid workers earning much less than those at sites that were not Fairtrade certified. Their examination of working conditions suggested similar disparities. With some variation, workers in the areas without Fairtrade-certified producers often fared better. And when researchers went back two years later, the gap had grown, meaning the laborers in areas with Fairtrade were even worse off on average.[60]

This is only one study and, like all research, it has meaningful limitations.[61] And although much of the research suggests the impact of Fairtrade can be mixed, there are indications that good can come from Fairtrade and other efforts to bring about change within the current system. Nonetheless, this type of work also points to the incredible challenge of trying to promote more sustainable practices and better working conditions through long and complex supply chains. It is costly to pay workers more, provide safe and respectful working conditions, and take the types of precautions that can reduce adverse environmental impact. This creates a strong economic incentive for gamesmanship, and the structure of the middleman economy can make it all too easy to blind consumers to the truth of what lies behind the goods they are consuming.

GREEN OR GREENWASHING?

These issues are also becoming more pressing in finance. Investors in Europe and North America are pouring money into funds that purport to prioritize Environmental, Social, and Governance (ESG) objectives alongside offering attractive financial returns. In the United States, the value of assets in such funds quadrupled between 2017 and

2020 and now represents a full 20 percent of the assets managed by index funds.[62]

One financial instrument that has become particularly popular in efforts to promote sustainability are "green bonds." As the name reflects, green bonds are meant to finance projects with environmental benefits. They are designed to appeal to investors concerned about the rapid rate of environmental degradation attributable to climate change, from rising sea levels and temperatures to stronger hurricanes and more frequent droughts and wildfires.[63] The issuance of these instruments has increased dramatically in recent years, surpassing $250 billion in 2019 alone.[64] This inspired researchers at the Bank for International Settlements to try to assess, quantitively, how issuing a green bond affects the "greenness" of the company issuing it. As they note: "A key question for investors in green bonds and similar instruments is how to verify that the promised environmental benefits are, in fact, delivered."[65]

The researchers began by identifying all bond issuances deemed "green" by one of four major providers of data on green bonds. They found that the proceeds from most of the bond issuances complied with the applicable standards, such as restrictions on the types of projects and activities for which the funds could be used and ongoing reporting. But when they expanded their analysis to look at the carbon footprint of the companies that issued the bonds, they found no evidence that issuing green bonds meaningfully reduced a company's total use of carbon a few years later. This suggests that even though companies used the additional proceeds in "green" ways, many simultaneously took other actions that offset those benefits. This type of research casts grave doubt on whether green bonds—at least in the early years—have on net had a meaningful, positive impact on carbon consumption even at the firms issuing them.[66]

A 2020 report from the U.S. Government Accountability Office highlights further problems with ESG ratings.[67] The GAO found that even as investors poured money into ESG funds, neither the SEC nor any other regulator provided clear, consistent rules regarding what companies must disclose and in what form. A half-dozen nongovern-mental organizations have stepped in to try to fill this void, but the GAO found a lack of consistency and convergence. The GAO further found that even when companies purported to use the same frame-work, they often still provided divergent qualitative and quantitative disclosures. In other words, despite the significant fees that investors pay to fund companies for supposedly greener and more sustainable investments, it was far from clear what they were actually getting in return. The burgeoning academic literature on the efficacy and infor-mational content of ESG ratings paints a similarly mixed picture.[68]

Again, the core challenge is that when investors or consumers are willing to pay a premium, middlemen and other companies have a lot of incentives to make it appear as though they should qualify for that premium. But so long as the chain separating investors from the underlying companies and projects their money is helping to fund remains long and complex, it is all too common for companies to game the system. And middlemen, such as the fund companies that purport to invest only in companies committed to sustainability or other aims, often profit from enabling this subterfuge.

There is still value in many certification schemes, and I often still seek out Fairtrade, organic, and other products that bear some certi-fication suggesting that they are better for the land or workers than alternatives. But the limited comfort I find in those labels pales in comparison to what I enjoy when I step outside the middleman game entirely and connect with farmers, makers, and others. Although there are far more tales that could be told about how and where the middle-man economy runs amok, let us turn away from what is going wrong and toward what is going right.

PART IV

DIRECT AND THE
PATH FORWARD

CHAPTER 9

CONNECTIONS, LOCAL AND GLOBAL

ONE COLD DECEMBER evening, I ventured to Jersey City to attend a "yoga and beer" session my cousin was leading at a local brewery. As I emerged from the underground train station, I found myself surrounded by outdoor vendors selling food and crafts. I was soon chatting away with an artisan selling Christmas ornaments made of bark, walnut shells, and other natural materials. I learned she was from Argentina and about the special wood she brought back for her creations.

I picked out a few ornaments to give as gifts. Then, wanting to do more for this woman who was so far from the warmth of her homeland, I also chose a small crèche of Mary, Joseph, and their infant son. As I was paying, she asked me to choose an angel to hang over the blessed family. I did. In turn, I asked her to keep the change from the cash I had handed her. She protested. As I started to push back, she explained that the angels were special to her. She wanted me to have it *as a gift*. I stopped insisting, nodded in gratitude, and walked away. I enjoyed a warm glow from within for the rest of the evening.

Lewis Hyde, in his classic work, *The Gift*,[1] explains the reason for that warm glow: "It is the cardinal difference between gift and commodity exchange that a gift establishes a feeling-bond between two people, while the sale of a commodity leaves no necessary connection."[2] I get to enjoy that glow again each holiday season, as I pull out the crèche and hang the angel over it. I recall that chilly evening and the woman's smiling face, and the warmth inevitably returns to my chest.

Gift dynamics are not present in every or even most instances of direct exchange; this chapter and the next also explore the many other, more practical benefits of going to the source. But the potential for gift-like dynamics to coexist with the commercial when an exchange is direct, and the connections and communities that result, bring to life how direct is about more than reducing the threats posed by the middleman economy. At its best, direct exchange can help foster a different type of society, one that rests on a recognition of our interdependence and the importance of our collective well-being rather than on assumptions of scarcity and individual advantage.

This chapter focuses on the heart of direct exchange. What it looks like, what benefits it can hold, the challenges it can bring, and how different the entire ecosystem can be when maker and consumer connect directly.

BEYOND LOCAL

Direct is often local by design, but the two need not go hand in hand. Wine, another consumption good, provides a great counterexample. Unlike fresh produce where local consumption is key to quality and taste, most wine has to travel. There are a limited number of places that have the weather, soil, and topography to produce grapes of sufficient character to make good wine. So some areas make a lot of wine while most regions don't make any.

The upside of this geographic concentration is that having a lot of good wineries clustered together can make for a popular travel destination. Visits to wine country can allow consumers to see the land where the grapes are grown and meet the people who help bring out the character in those grapes as they are transformed into wine.

I first discovered one of my favorite wineries, Navarro, when my sister and I spent a weekend near Anderson Valley, a couple of hours north of San Francisco. Anderson Valley is far less celebrated than Napa or Sonoma, and more reasonably priced — and better suited to my budget — because of that. I once read an interview with Ted Bennett, cofounder of Navarro, in which he explained that they sell a lot of wine to academics because professors "have an intellectual interest in wine, but don't have much money."[3] That's me in a nutshell.

Ted and his wife, Deborah Cahn, were among the first winemakers to set up shop in the area. In 1973, they bought a 900-acre sheep ranch and began planting grape vines. They started with Gewürztraminer, a white Alsatian varietal that they loved and that was hard to find in the States. It wasn't until the vines matured enough to yield grapes that could be fermented into wine that they realized it may be hard to sell such an unusual varietal.

Most wine sold in the United States goes through a three-tier distribution system, moving from winery to distributor to retailer before finally reaching a consumer. As Ted and Deborah learned the hard way, wine distributors in the late 1970s had little interest in investing the effort required to help consumers appreciate the virtues of a perfumey white with a strange German name. Ted and Deborah then tried going door-to-door to wine stores but still made no progress.

With lots of wine and no middleman wanting to help them sell it, Ted and Deborah set up shop at the winery. They cut down three trees, milled them right at the vineyard, and used them to build a tasting station. They also tried to sell the wine via mail order, starting with the Christmas card list maintained by Deborah's mother. Eventually, people started buying their wines, and then coming back to buy more.

Although it required far more effort in the early stages, this approach to selling their wine ultimately proved more profitable and sustainable than relying on distributors. To this day, Navarro offers free tastings at the winery, and they continue to use those tastings as an opportunity to connect with customers new and old. They also use mail order and a website to sell directly to customers that cannot make the trip.

One obvious advantage of cutting out two layers of middlemen, distributors and retailers, is not having to pay them. Although distributors and retailers cultivate opacity to make it difficult to assess precisely how much they earn, the markups and fees can be significant. According to some estimates, only half of the price paid for a bottle of wine typically goes to the winemaker. Many distributors also impose rules to ensure a hefty markup and to prevent wineries from selling their own wine at a lower price even when they sell directly to consumers.[4] These schemes ensure that a lot of the money Americans spend on wine goes to the middlemen. By contrast, when someone buys a bottle of Navarro wine, every cent goes straight to the winery's employees, operations, and founding family. This enables them to sell their wines at much lower prices than they would have to charge if sold in the traditional way.

Navarro also has a lot of fans. For example, Dan Dawson, onetime sommelier at the restaurant French Laundry, describes himself as "a disciple" and "humongous fan" of Navarro for twenty-five years.[5] Other people who know far more about wine than I do also see something special in Navarro's offerings, and regularly comment on how reasonably it is priced given the quality.[6]

This does not mean Navarro turns a profit every year. The winery has had years in the red. But it has consistently managed to survive, and even thrive. And even the hard years are partially a by-product of a conscious decision by Ted and Deborah to prioritize people over short-term profits.

For example, in contrast to most wineries, Navarro doesn't treat its agricultural workers as cheap, seasonal labor. Instead, it provides them year-round employment, offering them work on the bottling line, in shipping, or elsewhere. This limits Navarro's ability to cut costs during challenging years, but it also creates the stability that so many workers want and struggle to find these days.

Similarly, Ted and Deborah know that their long-term customers expect Navarro wines to taste a certain way, and they work hard to meet those expectations. When wildfires hit in 2008, for example, many wineries went ahead and used grapes tainted by smoky flavors for their regular offerings. Not Navarro. They sold the smokier grapes under a different label, "Indian Creek," and at a much lower price. They were also open with their customers about what they had done and why. This is just one example of how they have been both willing and able to prioritize integrity, even when it came at a cost.

Navarro also has a long-standing commitment to sustainability. The winery stopped using synthetic pesticides in the late 1970s when such decisions were rare, and it often seeks out ways to reduce its carbon footprint, such as using sheep in lieu of machines to keep the grass around the vines in check.

The ability to make these types of decisions is a reflection of how Ted and Deborah have approached growth: "really slowly." Ted and Deborah wanted to make enough money to send their kids, and now grandkids, to college, but as Deborah explained to me, "the whole motive is not monetary."[7] Never having any outside investors allowed them to think long-term rather than quarterly, and to make decisions that reflected their values and commitments to employees, customers, and family.

Navarro's customers have taken note. Patrick, a wine lover who lives in Southern California, first heard about Navarro in 1990 when a friend (and later winemaker) hosted a dinner featuring Alsatian wines. When Patrick asked whether any wineries in California made similar wines, he got a one-word answer: Navarro. Patrick immediately ordered his first few bottles and has been a loyal Navarro customer ever since. Patrick told me that he always feels good placing an order with

Navarro. He trusts their wine to be good and their pricing to be fair. He likes that when he visits, everyone at Navarro is down-to-earth and not condescending, like many in the wine world. He has visited so often that even their dogs are on a first-name basis.

I personally enjoy the detailed write-ups that Ted and Deborah prepare for each wine. These include facts, such as when the grapes were harvested, sugars at harvest, and pH, along with other reflections. They may discuss how excess rain in July affected the grapes or what a particular wine means to them. A recent write-up was titled "With a little help from our friends." In it, Ted and Deborah note that "mega-wineries with umpteen brands . . . virtually control wine distribution and the majority of small family wineries that blossomed in the last 40 years have been gobbled up by conglomerates." This makes them feel even more "fortunate" about their decision to "sell the wine directly to the consumer," helping them to remain independent.[8] The mix of information is far more extensive than what can fit on a label and more personal than anything a middleman could convey.

Information flows the other way as well. Through their myriad points of contact, Navarro gains timely and valuable insights into what their customers want and how shocks like the pandemic affect their preferences. The three-tier distribution system not only limits how other wineries communicate with their customers, it also limits how much they actually know about what customers are buying and thinking.

Ted is now in his eighties and Deborah is not far behind. Both of their children play leading roles at the winery, so Ted and Deborah could easily step down. Yet they keep working because they still enjoy and value the work. Their long-term customers are "almost like friends." They have had children of their agricultural team go off to college and come back and join management. And, as reflected in pictures of their grandkids helping with crush, time working is not necessarily time away from family.

A DIFFERENT APPROACH TO VALUE

Much of this discussion may seem unrelated to the fact that Navarro sells its wines directly to customers. The discussion has made little effort to disaggregate the customer experience from the experiences of the founders, workers, and others. Nor has it treated the mode of exchange as apart from the nature of the good, the processes through which it is made, or the impact of those processes on the people and places involved. This holistic approach stands in contrast to the economics-oriented framework used when we first looked at the middleman economy. This broader, more encompassing framing reflects the very different ways that people and goods are often treated in the two ecosystems.

In the middleman economy: People are often flattened into generic consumers whose top priority is getting the best deal and investors who want only to maximize their risk-adjusted returns; hyper-specialization changes the nature of work and leaves workers disconnected from the good they are helping to create and the people who will one day enjoy it; and collateral considerations such as environmental impact matter only to the extent mandated by law or justified by correspondent economic gains. The mode of distribution, how it enables power to accrue in middlemen and accountability to be lost amid long supply chains, and the state of the world today cannot be separated. They feed into each other in ways that were systematically, mistakenly, and often opportunistically overlooked by the excessive reliance on a particular type of economic framing. The seeds for the threats posed by the middleman economy were laid in the very foundations that gave rise to those structures.

Direct exchange, by contrast, tends to pull in the other direction, allowing—rather than negating—depth and multidimensionality. That Navarro has always sold its wines direct cannot be disaggregated from how it has evolved and how it operates today. Navarro's commitment to reasonable prices and consistent flavor profiles cannot be disentangled from its commitment to provide year-round employment to employees or its efforts to operate in a more sustainable fashion, and none of this can be separated from the distinctly close relationship between those making and those drinking Navarro wines. The notion that these different aspects of people and companies can be understood and addressed in isolation is one of the fictions of the middleman economy and modern capitalism that must be discarded to forge a better path.

GLOBAL AND CONNECTED

To fully realize this transformative power, however, direct exchange must go beyond the local. Local remains a centerpiece of direct exchange, but given that wealth remains far from evenly distributed across the country and around the globe, local exchange alone could affirm, rather than displace, existing hierarchies. Visits to wine country, comic conferences, and art fairs can help, but many people lack the time and money such direct tourism can entail.

This is where technology becomes critical.[9] One way technology can support direct exchange is by making it easier for makers to set up their own "shop." Instead of paying rent and investing in a physical storefront just to reach people nearby, today's entrepreneurs can set up a website and then ship their goods to customers across the country and around the world.

One entrepreneur who has done just this is Abena Boamah-Acheampong, founder of Hanahana Beauty. The company, based in Chicago, specializes in shea-based body butters, which it sells exclusively through its own website. All of its products are designed specifically for Black skin because they were created originally by Abena for use on her own skin.

Abena started making body butters in 2015, after she realized how little she actually knew about what went into the creams she was putting on her body. Having grown up in a home where her mother recommended shea butter for anything that ailed the skin, that is where she started. Abena's family roots go back to Ghana, one of a dozen African countries that produce most of the world's shea. She went online to learn about how to use shea for body butter, and kept searching until she found recipes created by Black women for Black skin. From there she experimented, mixing different oils to find blends that worked for her. She soon started sharing with family and friends, and their rave reviews helped her appreciate that she was onto something.

As Abena scaled up production, she wanted to source her shea directly from Ghana, so she got on a plane and went. Through a family connection, she found a driver, translator, and soon-to-be friend in Paa. He introduced her to the Katariga collective, one of a number of cooperatives of women who work together to transform raw shea nuts into shea butter. They showed her how they extract the seeds from their shells, and then all of the grinding, paste-making, and separation processes that follow. Without knowing why Abena was there, they gave her their time and knowledge.

When, a few days later, Abena sought to buy shea for her butters, she was so shocked by how little they charged that she paid double the price they quoted. Paying twice the going rate for shea is a practice Hanahana Beauty continues to this day.

Abena also made another commitment. She told them: "I think what you do is amazing. . . . And I'm going to make sure that people know about who you are." She realized how little most people do know, and how much they should, "about the people" behind the goods they use.[10] She started by editing video footage from that visit and posting it on Hanahana Beauty's website, where it remains freely available to anyone curious about how shea is made or the women who make it.

Abena has since deepened her relationship with the women of the Katariga collective. Some of the support she helps to provide takes a form that many would label charity. For example, Hanahana hosts a biannual health care day for the women where they can access screenings and basic care. Abena did this because the women in the collective specifically identified health care as an ongoing challenge, and she wanted to address their needs as they saw them.

Abena uses sales revenue to support these events, but she also created "The Hanahana Circle of Care" to enable customers to contribute directly. In 2020, for example, Hanahana raised $10,000 in donations to support the fall health care day for the women in the collective. Yet Hanahana doesn't have separate business and charitable arms, at least at this stage, nor does Abena see providing health care as charity. For her, it is about doing what is right, providing the pay and support needed to ensure that everyone involved in producing Hanahana's body butters can also care for themselves.

Social media has been key to Hanahana's success. This stirs up mixed feelings for Abena. She is private by nature and dislikes aspects of social media. But she also knows that she has been given the gift of a good eye and a knack for telling stories. This allows her to create content that resonates, helping spur interest in Hanahana and support for the Circle of Care. Hanahana's prices are far from cheap, reflecting the labor and care that goes into each jar. Yet Hanahana regularly sells out of its most popular products. This reflects well on both the quality of the products and the willingness of customers to pay more when they can see who benefits.

Social media has made it easier for Abena and Hanahana to participate in and help promote a global network of Black women. Hanahana's Instagram feed, for example, is full of quotations from Pearl Bailey, Toni Morrison, Alice Walker, Audre Lorde, Martin Luther King Jr., and others who have given voice to the Black experience. It is awash with inspiring images of Black women, sometimes alone, sometimes in community, sometimes in the United States, sometimes in Africa, sometimes laughing, sometimes serious—and yet always owning their space. To visit the feed is to be invited into a world where Black is beautiful, and self-care of Black skin is celebrated.

Like many Black-owned businesses, Hanahana enjoyed a surge in popularity when people took to the streets to protest the untimely deaths of George Floyd, Breonna Taylor, and so many others. The deaths spurred recognition of the ongoing violence too often perpetrated against Black men and women, just as Covid laid bare and accentuated massive structural inequities. Black and Hispanic Americans were more likely to contract Covid, lose their jobs, and suffer other adverse consequences than white Americans.[11] Yet Hanahana also reveals how many layers of impact are possible, but not always realized, in calls to "buy Black": Is it about a company's shareholders? Management team? Other employees? The respect shown to Black workers further up the supply chain? Hanahana's customers know who they are supporting at each of these levels.

Hanahana is a younger company than Navarro and it could still evolve significantly. Wherever the company goes from here, however, the Hanahana of 2021 embodies how direct exchange can transcend geographic and linguistic bounds to create new connections and move goods and money to where each is most needed without layers of middlemen. It also shows how disruptive and powerful a force that can be.

NO MIDDLEMEN IN SIGHT

For both Navarro and Hanahana, the decision to sell goods directly to their customers brings with it challenges. The informational and logistical hurdles that middlemen help bridge are real. Without the aid of middlemen, most of these burdens sit with Ted, Deborah, Abena, and their employees.

Gaining the attention and trust of potential customers is one of the biggest hurdles that makers face when they sell direct. For many of the "direct-to-consumer" companies that proliferated in the 2010s (explored in the next chapter), the high cost they pay to attract new customers is the biggest obstacle to long-term viability. And in contrast to DTC companies that could pay for expensive marketing and targeted Facebook ads, Navarro and Hanahana had to find a way to reach customers on their own.

The evolution of both companies, however, shows how having to overcome this hurdle without the aid of middlemen can be a blessing in disguise. It may have slowed growth, but the way those initial connections were forged makes them stronger. For Navarro, this comes through in the loyalty of their customers, in folks like Patrick who have been buying their wines for decades. For Hanahana, the depth of the connection comes through in the ability to raise funds for the Circle of Care. Not many companies can ask customers to make contributions to benefit workers up the supply chain and get that kind of response.

This level of customer loyalty also suggests something else: To go direct and survive usually requires a genuinely good product. Navarro's ability to convert free tastings into regular customers, for example, only works if people actually like the wine. For Hanahana, rave reviews from beauty editors—such as Courtney Higgs, in *Who What Wear*: "From the second my skin met its whipped shea butters, I've been a devotee. They're unreal!"—were key to expanding its visibility and it could not have proliferated at the rate it did if experts were not impressed by Hanahana's products.[12]

Logistics can also be tricky. A growing number of companies, like Shopify and Magento, ease these burdens by helping makers and other small businesses create and operate their own virtual storefront. In reducing the cost of "setting up shop," these types of infrastructure providers are critical contributors to the effort to expand direct exchange. As Shopify's CEO, Tobias Lütke, aptly explains: "Amazon is trying to build an empire, and Shopify is trying to arm the rebels."[13]

Yet many challenges remain. For wineries and other producers of alcoholic beverages, for example, an array of state laws and differing tax and licensing requirements can add to the difficulty, helping to explain why so many rely on middlemen. Getting goods into the hands of faraway consumers is also a hurdle. Small makers don't own fleets of trucks, nor do they often sell enough to negotiate preferential rates with big shipping companies.

Given the many challenges that arise when seeking to forgo

middlemen, it is not just makers that must exert some extra effort. Consumers bear some of the burden as well. They must be willing to try a new product even without a middleman to vouch for it. They have to accept that there are no Navarro wines or Hanahana body butters sitting in Amazon warehouses, ready to be shipped at no marginal cost with next-day delivery. Customers have to be ready to pay and to wait.

There may be some virtue in this. I know I could use more patience, and delivery on demand has likely helped erode the little that I naturally possess. And as we saw when examining CSAs, sometimes taking on the additional burdens that going direct can impose can create opportunities for growth and meaning making.

But not every inconvenience that arises from going direct is a blessing in disguise. There are ways that middlemen really do make our lives much easier. It takes work to find and buy goods without their aid. Similarly, it can feel good knowing that everyone involved in making a good was paid a living wage, but it often does result in higher prices relative to goods produced en masse overseas. Going direct can require meaningful investments of time and money and, even then, does not ensure a good match.

Yet, despite these many challenges, direct exchange is all around and growing in popularity. From the CSAs and farmers' markets that we first visited, to makers like Navarro and Hanahana, to the creatives selling their goods via Etsy and fund-raising via Kickstarter that we will visit in the next chapter, a lot of people are opting to take on the challenges that direct can entail.

A partial explanation for this rise comes from the many drawbacks of the middleman economy. As middlemen continue to grow in power and supply chains become ever more attenuated, the individual and collective benefits that arise from opting out of that system go up. Going direct allows consumers to know far more about what they are actually getting, can make it easier for them to express multiple values, and can even help save them money. From a systemic perspective, reducing the scale and scope of the middleman economy can reduce the costly inefficiencies that arise over time, increase resilience, and mitigate the other harms examined in Part III.

But these benefits are only part of the story. To transcend the limitations inherent in the still-dominant, economics-based mode of looking at the world, we need a different theory of value and exchange. We need a theory that, while also inevitably incomplete, illuminates a different set of dynamics.

This is why the chapter started with Lewis Hyde's work on gift economies and the story of the gifted angel. Hyde shows that in other places and times, gifts rather than commerce have been the dominant mode of exchange, enabling a different type of social and economic order. Yet *The Gift* has remained in print for nearly forty years because it also speaks to the way gift economies can coexist, albeit often uncomfortably, with market economies. And understanding gifts and gift economies can serve as that critical alternative lens for understanding the nature of value and the importance of how goods move from one person to the next.

GIFTS AND COMMERCE

Writing in 2019, Margaret Atwood described *The Gift* as "the one book I recommend without fail to aspiring writers and painters and musicians."[14] For Atwood, the book is about "the core nature of what it is that artists do," and the role of art in "our overwhelmingly commercial society. If you want to write, paint, sing, compose, act, or make films, read *The Gift*. It will help to keep you sane."

This is a powerful endorsement coming from a prolific author whose accolades include winning the Booker Prize, twice. More important, it bespeaks a very different understanding of the relationship among the act of creation, the mechanisms through which a creation changes hand, and the creation's value. Economics assumes sanity. It cannot provide a path to it.

For Hyde: "Scarcity and abundance have as much to do with the *form of exchange* as with how much material wealth is at hand."[15] This is a radical departure from the neoclassical view, in which scarcity is the default, and innovation and outsourcing create value. For Hyde, abundance is created not by producing "more" from less, but from the very act of giving and the human connection that results.

This is why the way goods move through an economy has such a profound impact on societal structures and well-being. For Hyde, the very act of commercial exchange has the effect of erecting or affirming boundaries. It is a form of exchange that perpetuates separation. Gifts do the opposite. "A gift, when it moves across a boundary, either stops being a gift or else abolishes the boundary."[16] And in a gift economy, goods flow not to those who can pay for them, but to those who need them.

In this way of seeing the world, separations and connections are not fixed but fluid. How goods move today determines tomorrow's communities and divisions. This lens suggests that the way the middleman economy disaggregates making into a multi-nodal process is a shift that results in more boundaries, more separation, and far more isolation relative to the time when people worked together in one place to produce a quilt, pin, car, or other good. The process of separating, not specialization, is the linchpin that matters.

As one might expect if this view holds merit, loneliness has increased alongside the rise of the middleman economy. Many public health experts see the loneliness epidemic that now plagues the United States as on par with the threat posed by obesity. A survey of 10,000 working adults conducted in the summer of 2019 found that more than three out of every five Americans is lonely.[17] Former U.S. surgeon general Dr. Vivek Murthy observed, "when I was traveling to communities across the country I found that loneliness was a profound issue that was affecting people of all ages and socioeconomic background."[18] As he explains, the data shows that "loneliness is associated with a greater risk of heart disease, depression, anxiety and dementia," in addition to being "associated with a reduction of life span."[19] Loneliness also "limits creativity" and "impairs other aspects of executive function, such as decision-making."

An economist might point out that the simultaneous rise of two developments is not proof that one caused the other. And so far as I know of, no study yet has shown a causal relationship between the rise of commercial exchange at the heart of the middleman economy and the spread of loneliness in the way that the spread of Walmart superstores has been shown to contribute to obesity.

Yet even leading economists recognize the limits of markets and marked-based interactions to promote human flourishing. For example, University of Chicago economist Raghuram Rajan, in his book *The Third Pillar: How Markets and the State Leave the Community Behind*, argues that markets (and, according to him, the state) have grown too big and too powerful relative to community. For him, community must be revived as the critical third pillar because of the distinct role it can play helping people develop a sense of identity, protecting members from feeling adrift, ensuring members are provided for in times of need, and encouraging the type of cooperation and commitments that cannot be easily reduced to contract or enforced in court.[20] Although Rajan is a financial economist whereas Hyde is a poet and essayist, their work points to similar deficiencies in today's social and economic systems. Both are focused on the limits of what markets can provide, and what gets squeezed out as markets and market-based ways of structuring interactions become too dominant. They are both attuned to the human need for connection and community.

Hyde's framework opens up new possibilities that go beyond fostering the type of community that even economists like Rajan see as vital. In exposing value that is degraded or ignored by the workings of the market, Hyde also shows the importance of the relational and meaning-making dynamics that get systematically undervalued in a capitalist system that seeks to put a price on everything. Embracing his lens allows new insights into why direct exchange has such a vital role to play in moving us from where we are to where many of us want to be.

GIFTS, DIRECT AND COMMUNITY

The CSA at Genesis Farm is a living example of how gifts can cultivate community and connection even when intertwined with commerce. The core exchange is commercial: I make a payment that gives me a legal right to a share of the vegetables, fruits, and flowers that the farm grows over the course of a season.

Yet the CSA works only because of additional understandings layered on top of that core exchange. One example is the "share table." If I am allotted more eggplants than my family will eat, I still take my allocation, but I then place the extras on a separate "share table." This table often also holds "gifts" from the farm, produce not quite up to their standards but sufficiently delicious and nutritious that someone should enjoy it. Everyone knows that anything on the share table is free for the taking. This system ensures fresh food goes home with those who are most likely to use it.

The gift exchange here involves a complex web of transfers. It is a variation of what Hyde deems a "circle"—a continuous flow creating connections and then community. The goods flow from those who have too much of something to those who could use a little more and enhance the sense of community and connection in the process.

Many CSA members give and receive in other ways as well. They participate in the annual carrot harvest, contribute to the auction, share poems or music at open mic nights, or help cook food for one of the many events the CSA hosts each year. Others accept these gifts, eating the delicious food, dancing to the music, or partaking in a freely offered craft project for their kids. People who give one day may receive the next. Repeated over time, in different forms but with traditions creating continuity, deeper bonds and true community form.

In the case of CSAs and many forms of direct exchange, the

exchange is in-person and local. In these spaces, direct overlaps with the "buy local" movement, which similarly recognizes that the commercial and personal/gift/relational can no longer be treated as entirely separate spheres. As Hyde noted when we spoke, "local exchange, even if it's cash exchange, does reproduce some of the intimacy of gift exchange . . . there's a sense that you know the merchant."[21] Direct takes this one step further by protecting the underlying good from ever having to ever enter a purely commercial stream. Spatial proximity also makes it easier for exchanges that have both gift and commercial components to contribute to a web of other gift-type interactions, fostering the type of community that Rajan identifies as vital and too often missing.

But in a world where separation feeds hierarchy, there is a need to break down barriers between neighborhoods, not just between neighbors. Hanahana Beauty reveals some of the ways that this is already happening. Gifts lie at the root of the company's origin story. When Abena first met the women of the Katariga collective, they gave her the gift of knowledge, not realizing she was a potential customer. She gave them a gift in paying double the price they asked. She captured and shared the gift of their story, posting the video as a gift to anyone with an Internet connection. Over time, the "Circle of Care"—a term that Abena chose without any knowledge of Hyde—grew. Hanahana Beauty continues to provide free and informative content. It continues to pay twice the going rate. Many Hanahana customers continue to offer "gifts" to help ensure the women in Ghana have access to health care and more.

As the group grows, people give to different people and receive from different people, breaking down barriers, forging connections, and transforming giver, recipient, and good in the process. This hasn't happened by chance. For Abena, community and connection have always been at the core of Hanahana. She sees Hanahana in community with the full Chicago-based team that makes, markets, and ships Hanahana products; she sees Hanahana in community with the women of the collective; she sees Hanahana in community with its customers;

and, she encourages movement and flows between these communities. Someone who is a customer one day may become an employee the next. Each of these communities feeds her core focus, cultivating a community of "Black people globally."

Like many businesses, Hanahana spoke out in 2020. When two police officers involved in the killing of Breonna Taylor were let off scot-free, Hanahana described the decision as "a reminder that our system doesn't value Black women," and it extended "hearts, thoughts + prayers . . . to Breonna's family and friends."[22] But in contrast to many of the large corporations that embraced Black Lives Matter that summer, Hanahana had been embodying and espousing that message for years.

GIVING, GIFTING, AND DOMAIN SEPARATION

This understanding of the power of gifts provides an opportunity to revisit my experience connecting with other parents of children born with unusual hearts. The problems posed by the middleman economy are not limited to for-profit corporations; they have also seeped into how people give. Layers of nonprofits are not uncommon, contributing to separations between giver and recipient. Efforts by established nonprofits to perpetuate their own existence are similarly common, even when such efforts may not be the best way to further the aims for which the organization was formed. There is also a difference between giving and gifting. Many motivations for giving to a nonprofit, whether to achieve status or perpetuate a self-image as different from those in need, are as much about separation as they are about connection, and many nonprofits facilitate this transactional approach to giving. This book does not delve deep into the world of giving in the way that would be needed to translate all of the ideas presented here onto the world of nonprofits and modern modes of philanthropy, but there are parallels that merit attention.

Like Hanahana Beauty, GoFundMe is designed to do more than

facilitate a quid pro quo. Conscious cultivation of connection and community are embedded in its very structure. The site puts the personal stories and updates at the center, and makes it easy for both givers and recipients to communicate both on and off the platform. This is what enables the connection and understanding that often make giving on GoFundMe feel meaningful. And consistent with how a gift economy is meant to work, gifts made via both organizations move money from people who, at the moment, are okay without it toward people who, at the moment, could really use it.

There are risks in blurring lines between philanthropy and business, between gift and commercial exchange, and between gift economies and market-based ones. As reflected in the ways middlemen have sought to greenwash everything from jeans to bonds, claims that commerce can be used to transform rather than affirm the structural problems embedded in today's economy are often little more than self-serving efforts to appease guilt without bringing about more fundamental change.

Yet, there is also a danger in taking the opposite approach—in assuming that work that doesn't feed the soul is inevitable and that commercial exchange can always and only be that. Direct exchange can promote connection and community, rework hierarchies, create meaningful and self-directed professional opportunities, and reduce loneliness. It can promote the distinct sense of abundance that comes not from what one owns, but from using exchange to connect with fellow humans.

Understanding gifts and gift economies illuminates the heights of what direct exchange makes possible. It also reveals the flaws inherent in the type of reasoning that enabled the growth of the middleman economy. Both lessons can help forge the path toward a better tomorrow.

CHAPTER 10

ALMOST-DIRECT, QUASI-DIRECT, AND THE LIMITS OF DIRECT

N THE MID-2000S, Silicon Valley got into the business of displacing banks. Their aim was to harness technology to cut out the middleman and facilitate more direct investment and exchange.[1] One prominent example was peer-to-peer, or P2P, lending. As a newspaper explained in 2007, the then-leading P2P lender, Prosper, "aspires to do for money what eBay did for your grandmother's teapot collection— create a person-to-person marketplace for consumer loans, and in the process, turn average people into bankers."[2] Prosper and others did this by creating platforms that allowed would-be borrowers to tell their story and explain why they needed the funds and enabled would-be lenders to select which borrowers to trust with their money. By harnessing the "wisdom of crowds," these platforms aspired to make loans available to more people, on better terms, while also creating a new type of investment.

Many embraced the notion that P2P offered a more personal and connected experience. An early account of P2P opened with the story of Colin Nash, a thirty-five-year-old "struggling with $12,000 in credit card debt," and Michael Fisher, who, at twenty-four, "was looking for a new investment."[3] Although the two never met, each helped the other get what they wanted when "Fisher loaned Nash $200" through Prosper.com. For both, the appeal of Prosper went "beyond the bottom line." "Photos and personal narratives" and other personalizing touches meaningfully enhanced the experience for Colin, Michael, and other users.

At first glance, P2P lending seems to be precisely the type of direct exchange this book identifies as vital for a better future. It cuts out an established and costly middleman, facilitates a type of connection, and could provide a social good by creating a lending regime less reliant on imperfect standardized metrics, such as credit scores.

The subsequent reality of P2P, however, bears little resemblance to these early, idealistic accounts. It turns out that individuals are very bad at determining whether a stranger is creditworthy, so early P2P loans performed miserably.[4] Additionally, research examining which borrowers got P2P loans and on what terms revealed significant bias.[5] If you were Black, it was held against you. Old or overweight? Not as significant, but also a negative. By contrast, mention military service and your odds of getting funded go up relative to other prospective borrowers with similar credit profiles.

Moreover, P2P platforms have not proven any more willing and able to help people when they really need it than banks. For example, when the pandemic hit and many people lost jobs and needed cash, loan applications at P2P lenders went up. But rather than meeting this heightened need, Prosper and other P2P platforms pulled up the ladder.[6] They started making fewer loans, and making those loans only to borrowers with high credit scores and verifiable income—the same people who could also get a loan from a bank.[7]

P2P platforms were able to quickly tighten lending standards because they had long ago abandoned a true peer-to-peer structure in favor of a "marketplace lending" or "digital lending" model. They still provide a nonbank alternative, but they increasingly rely largely on objective, verifiable data, just like banks. And, to gain scale, they have eschewed money from people like Fisher in favor of funding from private equity funds, hedge funds, and banks, sometimes via securitization. Adding to the complexity, regulators forbade the original structure in which funds flowed directly from individuals to borrowers. As a result, rather than being "direct," these loans now entail multiple layers of middlemen.[8] More recently, Lending Club decided that the best way for it to thrive is to become regulated like a bank or to buy a bank.[9] The direct alternative thus became the very middleman it once sought to displace.

There were early signs that P2P might not be able to realize the lofty aims set out for it. But to see those red flags for what they were, it is helpful to put P2P in context. That requires understanding two trends that sit alongside the recent rise in true direct transacting: a rise in venture capital–funded "direct-to-consumer" (DTC) companies, such as Warby Parker, Casper, Rothys, Dollar Shave Club, and Glossier, and a growth of marketplace platforms, such as Etsy and Kickstarter, that allow entrepreneurs and creatives, on the one hand, to find funding and customers, on the other hand. This chapter examines how these trends reflect and help address the problems posed by today's middleman economy. Yet it also explores why and how these arrangements fall short relative to quintessential direct exchange. Both understandings help explain why P2P was unlikely to succeed and provide important lessons for efforts to build a more direct, connected, and resilient economy.

THE DTC DECADE

If you spent any time on social media in the 2010s, you are probably familiar with the DTC trend. You may have noticed paid advertisements for Allbirds shoes, Away luggage, Harry's razors, ThirdLove bras, or Our Place kitchenware. You may well recognize their "specific visual brand identity (the now ubiquitous 'blanding') that favored sans-serif type, pastel color palettes, and scalable logos that were easily adapted to a variety of digital media," as explained by Leonard Schlesinger and co-authors.[10] Paid influencers accentuated the proliferation of posts raving about these companies and their products.

As Lawrence Ingrassia details in his book *Billion Dollar Brand Club: How Dollar Shave Club, Warby Parker, and Other Disruptors Are Remaking What We Buy*,[11] these companies were the darlings of the venture capital crowd during this period. Dollar Shave Club and Warby Parker were two early success stories. Both started online-only, each specialized in just one type of product, and each offered an alternative to an inefficient, established mode of intermediation. Prior to Dollar Shave Club, most razors sold by drugstores and other middlemen came from one manufacturer, Gillette. This type of arrangement can benefit both the dominant manufacturer and middlemen by enabling large profit margins that enrich both at the expense of consumers. Similar dynamics were at play in the market for eyeglasses that Warby Parker disrupted: A dominant manufacturer, Luxottica, enjoyed significant margins, enabling it to generously compensate the middlemen (often optometrists) it didn't control. Just as with razors, this enabled an equilibrium that had been stable for too long, as consumers overpaid, and Luxottica and the middlemen reaped the benefits.

Soon venture capitalists were pouring money into start-ups aspiring to disrupt middlemen in other domains. With this funding, entrepreneurs started selling shoes, couches, toothbrushes, pots, undies, luggage, and a whole lot more "direct to consumer."

Many of the companies that succeeded in the DTC decade did so because they grew into a domain that had been forced into stagnation by the middleman economy. As middlemen use their power to shape the playing field to their own advantage, inefficiencies grow and innovation lags. This made it easier for DTC companies to offer a superior alternative.

Consumers' growing demands for more sustainable modes of production and more connection to the people and places behind their purchases aided the DTC movement. These are not features that can be readily tacked on to long supply chains that evolved to make goods in the cheapest way possible. DTC companies showed that these goals could more readily be accomplished by starting fresh and creating short supply chains specifically designed to further such aims.

Clothing company Everlane exemplifies both the advantages and the limitations of DTC as a means to enhancing accountability. Everlane claims three foundational commitments: "Exceptional quality. Ethical factories. Radical transparency." This means a customer who wants to know more about the Cashmere Crew, sold for $100 in a dozen colors, is not limited to learning about it through the nearly 5,000 customer reviews (averaging 4.74) and product description, of the kind available on Amazon.

That customer can also learn just how much of the price she pays goes to what. A handy infographic shows that the company's "true cost" of making each sweater is $46.21: $30.35 for materials, $1.60 for hardware, $12.00 for labor, $1.76 for duties, and $0.50 for transportation. She can also learn that Everlane's cashmere collection is assembled in Dongguan, China, at a factory owned by two longtime friends, Mr. Chu and Mr. Lee. The cashmere comes from inner Mongolia and is woven into "yarn on Italian machines in Ningbo, China" before reaching Dongguan, where it is dyed, woven, sewn, and finished. In the accompanying pictures, the Dongguan factory, built in 2000, appears to be safe and modern, even if not particularly inspiring. Everlane notes that on a visit to the factory, they "toured the local cafeteria" and heard from the workers about how they spend weekends "run[ning] into town for movies or play[ing] mahjong in the courtyard."[12]

Everlane also embodies other benefits of the DTC model. According to Everlane, the cashmere sweater it sells for $100 would cost $230 if sold through a traditional retail middleman. Cutting out the middleman gives the company complete control over the virtual and physical environments in which their customers shop, creating a more cohesive consumer experience that can add to the sense of connection. And any questions or complaints go straight to an Everlane employee, enabling the company to identify challenges and opportunities more quickly and allowing it to use the response as an opportunity to further connect with its customers.

Like many true direct companies, Everlane makes additional efforts to operate responsibly. The profits earned each Black Friday, for example, go to the people employed in their factories, and the company is trying to eliminate virgin plastics from its supply chain in furtherance of its commitment to sustainability.

This model is working quite well for Everlane and its investors. It has earned an array of accolades and has been raking in sales, purportedly earning $50 million in revenue within five years of its founding.[13] Moreover, most of its items have thousands of glowing reviews. By any metric, the company is a success and going directly to customers has been crucial to that success.

On closer inspection, however, the glow around Everlane begins to fade. In March 2020, after months of coalition building, a cohort of the company's remote customer service representatives sent a letter to the CEO announcing widespread support to unionize and asking the company to voluntarily recognize the union. Four days later, 42 of the 57 remote customer service representatives were fired. According to the union, not a single employee who had voiced public support for the union was retained. Soon Bernie Sanders jumped into the fray, tweeting: "Using this health and economic crisis to union bust is morally unacceptable . . . I'm calling on @Everlane to bring workers back on payroll and recognize the @EverlaneU [Everlane Union]."[14] Regardless of the company's actual rationale, widespread layoffs in the midst of a pandemic do not evince a deep dedication to employee well-being, and stand in contrast to Navarro's approach of eating the occasional loss to provide stability to its employees.

Soon thereafter, management faced accusations of anti-Black behavior and otherwise failing to treat employees with respect. An internal investigation in the summer of 2020 found support for many of these allegations. The behavior at issue was not outrageous, but it did suggest a meaningful incongruence between the company's carefully cultivated image and the underlying reality. Similar tensions arose between the company's outward claim that it was committed to serving women of all shapes and sizes and internal reports suggesting it intentionally dragged its feet in expanding to plus sizes. According to one former employee: "Everything at the company at that time had to be aspirational" and "it was not aspirational to be fat."[15]

In 2019, Good On You—a website that rates the ecological and human impact of shoe and clothing makers—gave Everlane a measly "not good enough," effectively two out of five stars, and identified numerous deficiencies in the company's efforts to protect the environment and factory workers abroad. Closer examination of the company's disclosures reveals that alongside all of the information that is disclosed, a lot is not. The first names that might allow someone to identify the Mr. Lee and Mr. Chu who own the Dongguan factory, how well the factory employees are paid, and whether they enjoy any sick leave or other worker protections are among the many questions left unanswered. Nor is the company above playing on behavioral weaknesses, such as offering "free shipping for orders over $75" and giving customers a $25 bounty each time they get a friend to buy something from Everlane, in order to increase sales.

This is not to degrade Everlane's social and sustainability commitments, or the amount of information it provides customers, particularly in comparison to what happens in the middleman economy. A sales clerk at a department store couldn't provide a fraction of the information that Everlane makes available. But there is a gap between the reality of Everlane and direct at its best.

A close look at other DTC wunderkinds reveals similar dynamics. Many have shorter, more ethical, and more sustainable supply chains than traditional producers, and they provide customers more information about the people and places behind their goods. DTC companies also get more information from their customers, which they sometimes use to make modifications or expand offerings in ways that serve both the DTC company and its customers. For example,

Jeff Raider, famous in the DTC world for cofounding both Warby Parker and Harry's razors, requires all employees who join Harry's to spend time fielding customer service calls. He even does the same himself. He appreciates that there is no better way to serve customers than to understand them, and one of the benefits of the DTC model is the ability to engage directly with those customers.[16]

Avoiding traditional middlemen can also yield meaningful cost savings. And, because they specialize in just one product and depend so much on social media, reputation, and word of mouth, most DTC companies seem to create pretty good products. When a writer for *New York* magazine tried out bras from seven DTC bra makers, for example, she "was pleasantly surprised that all seven bras were, overall, comfortable, well-fitting styles that I'd be happy to wear any day."[17] Overall, the rapid rise of DTC is a good road map to the problems embedded in today's middleman economy, and their success shows the benefits of starting something new and more direct as a way of avoiding those problems.

Yet, a striking number of DTC companies have been subject to scorching exposés. DTC luggage maker Away was accused of fostering a "culture of intimidation and constant surveillance." Workers at ThirdLove bras complained of a "condescending" and "bullying" male boss despite the company holding itself out to the public as "by women, for women."[18] Dollar Shave Club, which helped spark the DTC revolution, was revealed not to be a true direct seller at all, but a repackager of razors available on Amazon and elsewhere.[19] Its closest DTC competitor, Harry's, now sells its razors via traditional middlemen, including Target. Many other DTC companies have succumbed to the idea that expensive ads on Facebook and other sites are the only way to attract customers, effectively allowing the tech giants to become a new type of middleman. Moreover, despite the supposed cost savings enabled by cutting out middlemen, many DTC offerings are priced well above what most people can afford. Some of this may reflect the true cost of producing goods in an ethical way, but it is also a reminder that these companies exist to make money and often have a high spend outside of production costs, like those targeted ads.

That DTC so often is a mixed bag, even if a huge step in the right direction, can be traced to the core of their business models and funding structures. Direct at its best arises from a connection, and subsequent relationship, between producer and consumer. With most DTC companies, however, the depth of that bond is limited by an omnipresent third party: the VCs funding the company.

WHY CAPITAL MATTERS

To understand how venture capital (VC) financing shapes a company's operations, it is helpful to know something about corporate law. A common misconception is that the law requires companies to maximize profits in a way that meaningfully limits their ability to do what is right by the planet, workers, or society. As I teach my students at Columbia Law School, this is simply not the case. The law provides directors and managers significant discretion to think broadly about what is best for a company and all of those affected by its operations. In fact, a foundational principle in corporate law is that shareholders *cannot* tell a company's directors or managers what to do. If a shareholder doesn't like the way a company is being managed, his options are to sell his shares or try to elect a new board of directors.

Yet the law also gives companies a lot of flexibility to choose how they want to arrange their internal affairs. VCs use this flexibility to demand an exceptionally high degree of influence over how a company is run. As I saw firsthand while practicing corporate law in the Bay Area, VCs regularly demand board seats and other rights that give them far more capacity to influence a company's day-to-day operations than shareholders typically enjoy. Because VCs typically have a wealth of experience and expertise, and they want the company to succeed, much of this hands-on engagement can be quite helpful.

But there are some instances when the interests of VCs diverge from the interests of a company's founders, employees, and customers. Most of the time, VCs invest other people's money—they too are part middlemen—and charge very high fees to do so. The only way for VCs to keep justifying their fees and attract new investors is to provide a very high rate of return and to get investors' capital back to them in a timely manner. That many start-ups ultimately fail increases the pressure on VCs to earn an exceptionally high rate of return on the few companies that do succeed. The net result is that VCs often pressure companies to grow very, very quickly, even when that may not be the best approach for a particular company.

Understanding how much influence VCs have and how they exercise that control helps to explain why the former CEO of one venture-backed DTC company discourages other entrepreneurs from accepting VC funding. In his view, their focus on "growth at all costs" can be detrimental.[20] The contrast between the VC approach and an organization such as Navarro or a CSA could not be starker. Growth is not bad, but when it is rushed, it can force a company to compromise on other values. That so many DTC companies rely on so much VC funding helps to explain many of the shortcomings of these companies that are now coming into view.

VCs are great at spotting opportunities. That so much VC money has flowed into DTC companies is a testament to the inefficiencies and other drawbacks of the middleman economy. The success of so many DTC companies reflects limitations inherent in what the middleman economy can provide. Many consumers want products better suited to their needs and more positively impactful on the world. DTC firms are showing how shorter supply chains and direct sales can further these aims. So long as they rely on VC funding, however, most of these companies are "quasi-direct" at best—a big step in the right direction while still offering something less than true direct exchange.

PLATFORMS

The other big thing of the 2010s was the rise of platforms, such as GoFundMe. Platforms serve many of the economic functions long played by middlemen. Platforms such as eBay, Etsy, and Kickstarter help people with complementary desires to connect and otherwise reduce the many barriers that stand in the way of mutually beneficial exchange. Many also provide specific services akin to those provided by traditional middlemen, such as aggregating reviews and facilitating the safe flow of money from one party to another, so they are a type of middleman.

Yet there are significant differences between these new platforms and traditional middlemen. Goods often flow from maker to buyer without ever passing through the hands of a platform; key terms of an exchange, such as the price, are often determined by the parties, not the platform; and, rather than operating as a barrier that precludes the two sides from seeing one another, most platforms facilitate direct communication and connection. In these ways and others, many platforms are more middlemen-lite than full middlemen. And in exercising less power over transaction terms and enabling more direct communication, platforms can enable a lot of the benefits of going directly to the source. In theory, a well-designed platform can marry benefits of the middleman economy (scale at the point of connection, making it easier for shoppers to get what they want and easier for sellers to find buyers for their goods) with benefits of direct exchange (human-to-human connection and the capacity to support small-scale production).

Just how many of those benefits are realized depends on a platform's design and how much control it is willing to cede. The world of platforms is in many ways a microcosm of the broader themes in this book, with some allowing meaningful connection and others playing a far more formative role in shaping who connects with whom and limiting direct communications between them, just like a classic middleman. Thus, just as with DTC, the rapid rise of platforms is a sign of progress, and one that reflects shortcomings in the middleman economy, while often remaining one step shy of true direct exchange.

One of the first digital platforms to transform what people bought and from whom was eBay's platform for secondhand goods. For decades, people had signed baseballs, antique chairs, old handbags, and all kinds of unused goods sitting in closets and basements. They could sell these items to a thrift store, but they often received little when they did, even when some people might prize the good.

As we learned in connection with real estate, when a good is idiosyncratic, finding the right match matters a lot. One buyer may take far more joy from a good than other potential buyers. Getting the good into the hands of that buyer increases the amount of joy that an exchange creates and increases the price the seller will receive for the good. Thrift stores are not great at achieving the perfect match because of the limited number of people who will walk into the store and sift through its offerings. In contrast, a site such as eBay can offer many of the benefits provided by the MLS—functioning as a centralized hub that allows sellers to reach more buyers and making it easier for buyers to search and sort until they find just what they want.

Like the MLS, eBay has the potential to make everyone better off. As goods move from people who don't value them to people who do, aggregate happiness grows. Moreover, eBay's secondhand market does this just by facilitating the movement of goods that already exist. That's good for the environment.

Another similarity between eBay and the MLS is that each was the first dominant platform in its respective domain, enabling each to obtain a critical mass of both buyers and sellers, and then grow their ranks on both sides. This characteristic of eBay—its need to have a lot of buyers to attract sellers and a lot of sellers to attract buyers— makes it a two-sided market, just like the MLS. Not all two-sided markets are platforms. A print magazine, which typically relies on attracting sufficient readers to justify high-cost ads and sufficient advertisers to subsidize printing costs and distribution, functions as a two-sided market without being a platform. But recognizing that most platforms are two-sided markets can help us better understand their distinct business model.

First, network effects often result in there being a small number of winners, and sometimes just one. Because both sides want to be where the other already is, dominant platforms can be very hard to displace, so platforms are often even more powerful than other types of middlemen. They also have strong incentives to use whatever resources they have to maintain dominance because loss of market share is often cliff-like rather than incremental. This can enable outmoded, inferior, or overly greedy platforms to remain dominant even after better alternatives arise.[21]

Second, two-sided markets can sometimes provide one side a seemingly good deal (such as a cheap magazine subscription) because so much revenue is generated on the other side (advertising).[22] This puts the onus on the consumers, often, to realize when they are "paying" with their attention, data, or in other nonmonetary ways. It also means that the fees a platform imposes on sellers often end up being paid in significant part by buyers, albeit not in ways that are transparent to them. This doesn't mean that buyers may not find great deals on eBay, but—just as in real estate—buyers are far more affected by price structures that they don't see than many realize.

Third, and relatedly, sellers and platforms can operate collectively in ways that skew buyer decision making. For example, on Etsy, there is a fee to post an item, another fee when an item is sold, another fee paid to PayPal for processing the payment, and a range of optional fees to make a listing more prominent, such as having **the title of the item in bold** or having it appear under "related items" when shoppers peruse similar listings.[23] It takes an expert to navigate these trade-offs, and both buyers and sellers need guidance to avoid making mistakes in the process. In other words, eBay doesn't maximize profits by laying out the site to help each buyer get what he most wants with minimal effort. Rather, eBay uses its layout and fee structure to maximize the fees it can earn on each sale without scaring away too many buyers and sellers.

These insights into the workings of platforms provide a frame for looking at the diversity of platforms available. We have already learned a great deal about one dominant platform that exerts significant control and operates much like a traditional middleman in terms of the power it retains for itself: Amazon Marketplace, the platform it operates for third-party sellers. As we saw in earlier chapters, Amazon's decision to integrate this platform with Amazon's retail business proved pivotal in its ability to become the leading seller of so many types of goods. Its platform business has consistently grown faster than its traditional retail business, yet the growth of each feeds the other and provides Amazon data that allows it to be even more calculative in deciding what to sell itself, how to price those items, and what fees to impose on third-party sellers. Understanding the platform component of Amazon is critical to understanding how it amassed so much power so quickly, the degree of leverage it enjoys over third-party sellers, why so many people cannot imagine giving up shopping on Amazon for a year, and why it will be so difficult to weaken its dominance without government intervention.

Shifting from eBay in the opposite direction of Amazon reveals other platforms that operate quite differently. Two of the most successful and familiar are Etsy and Kickstarter. Etsy was started in Brooklyn in 2005. Its strategy to approaching the two-sided market challenge was to cater to artisans and other creatives, creating a platform that was friendly for them to use and shielding them from having to compete—on the site at least—with manufactured goods. This worked in attracting individual creators and shoppers that wanted the unique goods they had to offer. Etsy both benefited from and contributed to an increase in crafting and demand for handcrafted goods.[24]

In contrast to eBay, Etsy made community and connection a top priority. Sellers regularly cite their ability to connect directly with buyers as central to their success on Etsy and a reason they enjoy selling there.[25] Etsy also helped sellers connect with customers on other websites, such as Facebook, enabling these creators to tell even more of their story and often blurring the line between personal and professional. In its early days, Etsy also worked to foster a sense of community among its sellers. As former CEO and chair Chad Dickerson explained in 2011, a "team mentality has always been part of the spirit of the Etsy community,"[26] and it helped sellers band together in teams through which they could exchange feedback and encouragement.

Etsy also embodies how the right infrastructure can facilitate changes at both ends. Just like the middleman economy changed how things were produced and what people consumed, Etsy shows how making it easier for people to connect outside the middleman economy can have similar ripple effects. After I realized I could buy hand-crafted goods on Etsy and have them shipped directly to friends, for example, I found it easier to reduce my reliance on large middlemen when buying gifts. In this way, Etsy is the type of development that could slowly erode the power of dominant players like Amazon by providing consumers an outside option, and one they can access from their laptop no less.

Even more profound is the way Etsy created new opportunities on the opposite side of the exchange. As consumers came to use Etsy more and more to buy goods, this demand created new opportunities for sellers. There are numerous accounts of people who used Etsy to convert a hobby into a fulfilling job.[27] As of 2020, 87 percent of Etsy sellers identify as female, 80 percent are one-person businesses, and 43 percent are supporting other people, such as kids.[28] Etsy doesn't just allow sellers to find jobs that feel meaningful, it allows people who may not be in a position to take a traditional or in-person job to make money with the time they do have, and often without having to leave home. Here, too, Etsy creates the type of outside option that can help weaken, even if only modestly and indirectly, the power of the middlemen giants by giving people with a creative or entrepreneurial bent an alternative to Amazon. The progress thus far may be modest, but in light of evidence that lower wages and fewer outside employment opportunities are among the most significant harms that can flow from excessive concentration, the importance of these types of outside options is hard to overstate.[29]

Much has changed at Etsy in recent years. In 2013, Etsy started allowing sellers to rely on outside production assistance. This allowed greater scale to expand from just the point of intermediation to the production as well, leading to more goods being available at lower prices, but also disadvantaging those who were selling goods that they really did make by hand. In 2015, in part to enable VC investors to cash out on some of their investment, Etsy became a public corporation. This exposed the company to pressure from activist shareholders who believed the company could and should make profitability more of a priority. The result was the ouster of the company's beloved CEO Chad Dickerson, along with eighty other employees and further operational changes.

The new head of Etsy, a former eBay executive, has helped the company grow more quickly and become more profitable, but only by further deviating from its original commitments. The company abandoned its B Corp certification—a way companies can commit to prioritizing social and environmental goals alongside profitability.[30] Putting itself in league with Amazon, the company now prioritizes sellers that offer "free shipping," which many smaller sellers find difficult to provide and remain profitable. And despite having once prohibited the sale of manufactured products, the website is now replete with goods made via complex supply chains that the same seller also hawks on Amazon and elsewhere, so long as he designed it.

Amid all this change, much good remains. Etsy still enables almost-direct exchange; it still promotes community and communication; and it still provides both makers and consumers an opportunity to find customers and goods they could never otherwise reach. When Covid hit, Etsy's strengths shone bright. In contrast to the rigidities embedded in long supply chains, Etsy and its sellers could pivot quickly and easily. As demand for masks and homemade baked goods skyrocketed, and many people found themselves suddenly unemployed, Etsy helped to bridge these ends. Shoppers got ready access to a plethora of hand-sewn masks and fresh scones, and people without formal employment found a way to make money from the confines of their homes.

That Etsy has retained so much of "the good" despite the focus on profitability suggests even Etsy's new management and shareholders recognize that a lot of money can be made helping to connect small sellers and buyers and enabling the type of personal touch and connection that the middleman economy cannot replicate. The rapid ascent of Etsy, like the growth of DTC, is a sign of disillusionment with what the middleman economy has wrought and a recognition of the value that can be unleashed in facilitating better approaches to production and exchange.

P2P VS. KICKSTARTER

The forays into DTC and platforms help to explain why P2P, as a mode of connecting peers and facilitating financial flows among them, failed. Just like most DTC firms, Prosper, Lending Club, and early P2P lenders relied on VC funding. This gave them the capital they needed to grow, but also put them on a path where they had little choice but to grow rapidly, change, or die. Moreover, despite using a platform structure, the nature of the P2P exchange—cash today in exchange for an obligation to pay back even more cash tomorrow—is not one that requires a particularly distinct match, nor is it conducive to the personal touch that often accompanies an exchange on Etsy.

To better understand why P2P as a true peer-to-peer exchange failed, and what lessons that failure may hold, it's helpful to compare it to a successful financing platform, Kickstarter. Kickstarter holds itself out as "a new way to fund creative projects." It enables filmmakers, musicians, video game designers, and other creatives and entrepreneurs to seek funding for "projects, big and small." It has helped them finance projects ranging from theatrical productions to innovative bike racks and new lines of yoga clothing, and has often helped them garner valuable press, customers, and other connections in the process. A typical proposal features information about the proposed project, the creatives behind it, and an array of other material—such as pictures, videos, and stories—intended to capture the spirit of the proposal and engage with would-be funders.

Like the original paradigm for P2P lending, Kickstarter uses an Internet-based platform to enable people to raise and give funds; and, on both, a project must hit a minimum threshold level of support to be funded, thus harnessing the wisdom of a crowd. Yet, there are important differences between the two. The P2P exchange is at core financial and obligatory: Money flows one way today in exchange for an obligation that it flow back, with interest, in the future. As the performance of early loans showed, most people are not experts in credit risk.

On Kickstarter, by contrast, creators may offer goodies, including tickets to the show, early models of the good to be produced, or other paraphernalia in exchange for support, but promises of financial rewards are prohibited. Explicitly prohibiting loans, equity stakes, and other common financing devices changes the nature of the exchange, and, therefore, who is coming to the table and why. People do not back products on Kickstarter planning to get rich. They do so because they think an inventor has a good idea and they want what he has to offer or are inspired by an artistic project. They are, in short, harnessing more parts of who they are and providing information about what people want and what moves them in conjunction with providing financing.

This shapes who seeks funding via Kickstarter and why. Borrowers seeking funding via P2P just wanted a loan, ideally on better terms than they could get elsewhere. Creatives and entrepreneurs, by contrast, often use Kickstarter to test out ideas and cultivate connections with the very people they hope are interested in their good or other offering. Bob Frantz, a creator in Ohio, sees his Kickstarter campaigns as, in part, pre-sales of his graphic novels, allowing him to gauge interest and build a customer base in addition to raising the funds he needs to execute on an idea.

Bob is a stay-at-home father and a creator of podcasts and graphic novels. For him, Kickstarter has been a creative lifeline. He doesn't have the means to hire the artists and others needed to produce high-quality graphic novels on his own and he has yet to find a commercial publisher for his work. On Kickstarter, he and his co-creator have had two campaigns that brought in a little over $25,000, along with some smaller successes. They don't tend to make money after paying for the project and the various goods promised, but they raise enough to continue to create, to hire the experts needed to bring their visions to life, and to provide them a finished project that they can continue to sell at comic cons and elsewhere. Funding projects via Kickstarter also provides less tangible benefits. As Bob explained to me, "every one of your backers is validation, like someone believes in your project."[31]

Jess Baldwin, a music teacher, expressed similar feelings in her effort to raise $10,000 for studio time and professional musicians so she could produce her own high-quality album. In her first update, after raising more than $2,600 in the first day, Jess admitted to tearing up multiple times and noted that she had not been prepared for how "uplifted and supported" she would feel seeing so many people she knew, not to mention strangers, offer money to help make her dream a reality.

Bob also gives money on Kickstarter. Despite not having much cash to burn, he has provided some funding to more than 120 projects, some by friends and many by strangers. For Bob: "It feels wonderful, because I can help. . . . I can help someone else achieve their goals." As someone who can sit there and just stare at the screen while his campaigns are going, he knows how much each donation can mean. He has experienced from both sides the way early backers "feel like they're a part of the project" and "take pride" and even feel a sense of "ownership in your work."

And just as with true direct exchange, there are often positive spillover effects from the connections people forge on Kickstarter, even beyond the goodwill it can engender. Wharton professor Ethan

Mollick surveyed all 61,654 Kickstarter projects that got funding in excess of $1,000 before May 2015.[32] Based on responses from more than 10,000 of those creatives, he found that each dollar a creator raised on Kickstarter was correlated with $2.46 in additional revenue outside of Kickstarter. His work suggests that by 2015, Kickstarter had already played a role in the creation of more than 5,000 full-time jobs, 160,000 temporary positions, and more than 2,600 patent applications.

Like all platforms, Kickstarter is a middleman, and much of its success can be attributed to the way it helps creators and supporters forge connections that would not have been possible twenty years ago. Yet the nature of the connections it helps to forge and the design decisions it has made about which connections to encourage situate it toward the middleman-lite end of the spectrum. This allows it to provide more of the benefits associated with going to the source—and means it also has some of the challenges, like fewer middleman-assurances of quality of performance.

The contrasting fates of P2P and Kickstarter serve as a reminder that the benefits that direct exchange and shifts in that direction can offer are not equally helpful across all types of settings. One important difference between more direct and more intermediated forms of exchange is that direct makes it so much easier for people to come to the table as multidimensional beings seeking a multidimensional exchange. This is what lies at the core of Kickstarter. People who provide money are also revealing something about what they like, what excites them, and what leaves them cold, and that information provides value to creators apart from the sums transferred.

The same was never true for P2P, despite the nice stories. The exchange is at its core financial and little more, and there are some distinctly good reasons for intermediation in financing. The ability to get a lump of cash has a long history of motivating less-than-honest claims, enhancing the importance of intermediaries' expertise in weeding out fraud and assessing the probability someone will repay a loan. The process of collecting money when it is owed has never been a pleasant one, and is unlikely to be the type of interaction that fosters intimacy and goodwill. And the value of diversification means individuals providing funds are often better served by going broad rather than deep. P2P platforms recognized this, but efforts to ensure that lenders were exposed to lots of borrowers and that borrowers got funding from lots of lenders had the effect of further depersonalizing the exchange. This suggests financing may be a domain where a middleman, or even two, will often be helpful.

Similar dynamics also arise in other domains. It is not by chance that many of the stories of direct at its best involve goods that require relatively minimal processing. In many ways, this could be a distinct benefit of going direct: Less processed foods, for example, are often healthier than more processed ones. When buying a car or computer, however, a multi-stage production process is almost inevitable. The typical smartphone requires inputs from around the world.

The lessons contained herein remain relevant in such settings. For example, the computer company built by Michael Dell was the DTC phenom of its day, and cutting out an extra middleman was key to its success.[33] And the current DTC movement shows that there can be real gains from crafting shorter chains in ways that allow more communication, transparency, and accountability. Nonetheless, this does suggest that there will be domains where putting these lessons into practice will entail looking for ways to shorten supply chains and reduce layers of middlemen rather than eliminating them altogether.

One reason that this book has devoted so much effort to explaining what middlemen do is that it is very hard to disrupt an established class of middlemen without first appreciating the value they bring to the table. Today's intermediation schemes may be ridden with inefficiencies, but most arose for good reason. Understanding why middlemen exist in a domain is key to identifying the hurdles that must be overcome for a direct or almost-direct option to be viable. Only by understanding middlemen and the benefits of intermediation can we begin to identify the domains most ripe for change, and the conditions required for a more direct alternative to take hold.

A related implication is that direct exchange has the potential to be the most valuable and transformative when it allows the parties to harness some of the distinct advantages of going direct, whether that's more transparency around sourcing or an opportunity for a personal interaction. The range of circumstances where this may be valuable or possible is still evolving. When it comes to investing, for example, many people today may want nothing more than to maximize their risk-adjusted returns, and, on the other side, individuals and companies may be focused on the cost of funds raised or borrowed. But there is no reason to assume such preferences are fixed. The rise of the middleman economy and the assumptions enabling its spread haven't just affected policy making; they have also shaped how people see themselves when they don a hat such as consumer, investor, or borrower. Even if the ESG movement so far has neither lived up to its promises nor forced investors to make difficult trade-offs, the pace of its ascent suggests people may one day be willing to make different types of decisions, and different types of trade-offs. This could create opportunities—and demand—for ways to invest locally, more ethically, or in other ways that do bring more of people's individual preferences and values to the table.

The proliferation of DTC companies and the rise of new digital platforms have been positive developments. The rapid rise of both are testaments to just how much opportunity lies in the inefficiencies and other deficiencies embedded in the middleman economy. They also show how progress can be made in large-scale ways, and how even movements in the direction of direct exchange can be useful. Although a victory dance is premature, when put alongside the growth in true direct exchange, these developments show how much progress has been made and how much more remains possible.

CHAPTER 11

FIVE PRINCIPLES FOR POLICY MAKERS, COMPANIES, AND THE REST OF US

I HAVE DEVOTED MORE than a decade to studying the middleman economy. Throughout, I have regularly used my life as a testing ground for the insights that come from my research. Over time, this interplay has yielded five simple principles for navigating the threshold question of "through whom" to buy, invest, and give. These principles can be used by anyone, in the context of virtually any decision. This universality is possible because context matters so much in application: The principles both invite and require people to take their positions, preferences, and constraints into account when using them to guide a given decision.

The principles can be deployed by consumers looking to make more ethical decisions, managers trying to save money for their companies, and entrepreneurs seeking the next business opportunity. They can also be used by concerned citizens and policy makers who want to help shift power away from the middleman economy and back into the hands of the people who create and consume. Although each of us has the power to make different and better decisions, the middleman economy is too entrenched, and middlemen themselves too powerful, for individuals and organizations to transform the system without help. Federal, state, and local lawmakers all have a role to play in helping to promote a better balance. This chapter introduces the principles and then explores how individuals and businesses can use them to make better decisions and how they can also be used to illuminate the policy reforms needed to bring about meaningful change.

THE PRINCIPLES IN A NUTSHELL

Principle #1: Intermediation Matters: It is not just what we buy or to whom we give but also the structures through which we transact that matter. Whether a transaction is direct or entails layers of middlemen shapes the experience, the nature of the ultimate product or investment, and the ripple effects of the exchange. This may seem obvious to anyone who has read this far, but it can be easy to forget in the hustle of daily life. Thus, the critical first step is to recognize just how much is at stake in decisions to rely on middlemen, forgo them entirely, or choose more selectively among them.

 Principle #2: Shorter Is Better: The shorter the intermediation chain, the better. There is no single optimal length, but layers of middlemen often spell trouble. As we saw in Chapter 1, nearly four thousand people got sick, more than fifty died, and far more suffered psychological or economic damage as public health officials struggled to identify the origins of an *E. coli* outbreak in Europe. Similarly, the information gaps that exacerbated the 2008 financial crisis were the by-product of layers of investment vehicles—collateralized debt obligations, mortgage-backed securities, asset-backed commercial paper, money market mutual funds—that made it nearly impossible for anyone to know just how risks were allocated across the system. On a more banal level, despite having tried, I cannot figure out just where the nuts in my mixed nuts or the oats in my cereal were actually grown. Shorter supply chains enhance accountability, reduce fragility, and can sometimes lead to meaningful cost savings.

Principle #3: Direct Is Best: When an exchange is direct, both sides see and have the opportunity to know the other. Many of the resulting benefits are akin to those that come from cutting out excess middlemen: greater accountability and resilience, more positive ripple effects and fewer negative ones, and more gains to be shared between maker and consumer. Beyond these advantages, however, direct exchange can also set the stage for other, powerful dynamics: enabling connection, promoting community, counteracting the loneliness that remains so pervasive, and reworking hierarchies. Rather than reducing individuals to types, direct exchange allows people to come to the table as the multidimensional human beings that they are, and enables them to feel more human as a result. Direct exchange is not the right option all of the time, but it is a key ingredient to a balanced life and a healthier economy.

Principle #4: Follow the Fees: Given that middlemen are here to stay, it is important to know which ones to use and for what. Understanding how a middleman makes money can make it easier to detect the tricks middlemen often use to encourage customers to spend more or nudge them toward a higher fee product or investment. It can also illuminate which middlemen to trust. A neighborhood bookstore may be a middleman, but, like Navarro, its viability depends on the willingness of customers to return again and again, helping align its interests with its long-term customers. Homing in on how a particular middleman is compensated can go a long way toward better decision making.

Principle #5: Bridges Can Help: More direct exchange will likely mean more local trade, investment, and donations. This itself is not bad, as local has always been core to community. But if direct stops there, its capacity to transform the economy for the better will be limited. Today's world is not flat. Both the real and virtual worlds we inhabit are hierarchical and divided. For direct to help smooth structural inequities, it must go beyond deepening existing bonds. This can happen in a few ways. For one thing, what we see as "local" can evolve. During the pandemic, for example, many city dwellers left their urban confines to go apple picking or buy a Christmas tree.[1] This was often a new form of direct exchange and one that allowed them to appreciate the richness of the land not far from where they reside and connect with people who may have voted differently in the last election. More important, as embodied in the many different communities Abena promotes through Hanahana Beauty, community can take many forms, and can even span across and between continents. Consciously cultivated, direct exchange can play a role helping to disrupt embedded inequities.

PUTTING THE PRINCIPLES TO WORK: INDIVIDUALS AND ORGANIZATIONS

To really understand these principles, it is helpful to see how they can be put to use. What follows are some examples of how each principle can be put into practice by individuals, families, and organizations.

The examples illustrate that these are principles, not rules. Rules provide clarity and ease of execution, but they don't allow for nuance and individuality. Principles, on the other hand, provide users both the opportunity and obligation to take their distinct preferences and constraints into account in determining the right choice for them. There is no "right" answer to the question of when to go direct, when to use a middleman, or what middleman to use, but some options are better than others. The principles help to illuminate the trade-offs, enabling users to make decisions that are right for them.

#1: RECOGNIZE THAT INTERMEDIATION MATTERS

The threshold lesson is that taking time to consider "through whom" when buying, investing, or giving can pay off in spades. Relying on a large middleman is often the quickest and easiest choice, and sometimes it is the right choice. But pausing can reveal times when the obvious choice is not the best one.

When a good is bought through a middleman, it is often impossible to learn what portion of the moneys paid is going to that middleman or other middlemen along the production path, and how much is reaching the workers who grew the raw materials or labored to transform those materials into a finished product. Much of what is unknown is hidden for a reason. Alongside enabling efficiencies, long supply chains often facilitate the exploitation of workers, environmental degradation, and other efforts to take advantage of uneven regulatory protections across jurisdictional bounds. Each purchase also feeds the existing system, providing large middlemen the revenue, data, and market share that they use to perpetuate their dominance.

Just as important, looking past the obvious choice can reveal previously unnoticed opportunities—be they opportunities to connect, to express other values, to save money, or to have an adventure by going outside one's comfort zone. It can help to approach direct exchange and efforts to seek out shorter intermediation chains as experiments rather than commitments. Try something new, see how you feel about it, and then use those insights to shape future decisions.

Understanding the importance of intermediation can also help entrepreneurs identify new opportunities. The early DTCers recognized that they could create real value by displacing entrenched middlemen, but also by providing more transparency and accountability than the middleman economy affords. Other opportunities lie in becoming a better middleman. With respect to investment, for example, finding ways to use shorter chains to create investment opportunities that offer lower rates of return but promote sustainability, support local businesses, or further other aims could be both profitable and socially useful.

Many large companies are already in the process of reexamining the intermediation systems on which they rely. The supply chain disruptions triggered by the pandemic, for example, forced many companies to recognize the vulnerabilities that can arise from relying on long, disaggregated supply chains. Others are doing so in response to increasing consumer demand for more information about the social impact of the goods they consume or as a way of raising capital at a lower cost. Understanding how much is at stake in that threshold determination of "through whom" can set the agenda for established practices that merit renewed scrutiny.

2: SEEK OUT AND HELP CREATE SHORTER SUPPLY CHAINS

Reliance on middlemen is not an all-or-nothing choice. There is a lot of middle ground between buying corn directly from a local farm stand and the layers of middlemen behind the processed, corn-based cereal on the shelf at Walmart. Quite often, the best way to harness the lessons contained in this book is to strip out a layer or two of middlemen rather than going all the way to the source. Even these efforts can yield significant benefits in terms of reducing the opacity and lack of accountability that are the common by-products of overly long and complex supply chains.

For example, finance is an area where many people may still want to use a middleman much of the time. Index funds, such as those that are linked to the S&P 500 or the Russell 3000, for example, provide an easy and low-cost way of holding a diversified bundle of stocks. And there are times when particularly skilled middlemen can provide the expertise that individual investors lack while also providing companies the patient and informed capital that can help them succeed. Berkshire Hathaway, controlled by Warren Buffett, shows how this can work (although recent debates about the adequacy of Berkshire Hathaway's climate-related disclosures and its approach to governance also illustrate the importance of choosing the right middleman when taking this approach).[2]

But the number and structure of middlemen in finance goes far beyond what is needed or optimal. In 2020, for example, the Securities and Exchange Commission was forced to issue new rules for "funds of funds"—in which mutual funds hold other mutual funds—because assets invested in such arrangements had grown from $469 billion in 2008 to $2.54 trillion in 2019.[3] These types of layered arrangements result in more fees for the middlemen—here, fund companies—while doing little to make investors or society better off. There's no reason to accept more middlemen when fewer will do.

The auto loan market further illustrates how extra middlemen can increase costs. When seeking a loan to buy a car, consumers have two choices: go directly to a bank or other lender or get the loan through a car dealership that then arranges for a loan from a finance company.[4] Neither option is direct, but getting a loan from a lender cuts out one middleman—the car dealership—resulting in a shorter capital supply chain. Research shows that going directly to the lender tends to lead to lower financing costs for borrowers, as they avoid any chance of a dealer markup. Yet only about 20 percent of borrowers take this approach. The other 80 percent pay more as a result. One study found that consumers who financed loans through car dealerships in 2009 alone will collectively pay an additional $25.8 billion in interest as a result of that decision.[5]

Making matters worse, these costs are disproportionately borne by borrowers of color. One study found that Black car buyers who financed through a dealership were far more likely than white buyers to be charged a dealer markup, which increases the effective interest rate paid by the consumer.[6] Another study found that Black car buyers are more likely than white car buyers to finance their vehicle acquisition through a dealer, that Black borrowers paid, on average, a full percentage point higher interest rate than white borrowers, and that they were more likely to pay a very high interest rate.[7]

Experimental research helps explain why Black borrowers so often end up paying more when dealers, as an additional layer of middlemen, get involved. In one study, researchers sent pairs of testers (roughly matched by age and gender) into car dealerships in Virginia.[8] The white tester in each pair had a lower credit score (and usually a lower income) than the nonwhite tester. Nonetheless, the majority of the time, the white testers were offered more financing options and more favorable financing options than the nonwhite tester. On average, had the loans been entered into, the typical nonwhite tester would have paid an additional $2,700 over the life of the loan they were offered relative to their white counterparts. Cutting out excess middlemen is important for all of us, but given the realities of implicit and explicit bias, and the ways middlemen have so often exploited inequities, it may be particularly important for Blacks and other historically marginalized groups.

Cutting out a middleman can require work. For example, when ordering food from a restaurant, people often have the ability to choose between ordering directly from a restaurant or doing so through an app such as Grubhub or DoorDash. The apps are often easier to use, provide an array of options in one place, and can make it easier to have food delivered. Yet, those apps also charge high fees that can eat away at the thin profit margins of the restaurant actually making the food.[9] Each additional middleman that must be compensated means higher costs for the consumer, less revenue for the maker, or both. Occasionally ordering direct, even if it means a short walk or picking up the phone to place an order, can help local restaurants stay afloat.

For companies, shortening supply chains is one way to make progress on an agenda of reevaluating sourcing and delivery. Sam Walton learned early the value of cutting out middlemen farther up the supply chain. Even before he opened Walmart, when he was running a franchise, he sought every opportunity possible to source products directly from a supplier rather than buying them through the franchisor—an additional middleman that added costs he would rather avoid. It was a lesson he applied over and over again at Walmart.

Other times, cutting out an extra middleman requires fighting for legal change. As discussed in connection with Navarro, most wine travels through a costly three-tier distribution system. At this point, you won't be surprised to learn that distributors have worked hard to get states to adopt laws that protect their place in the system. Costco has filed numerous lawsuits to try to invalidate these rules.[10]

Costco knows that bypassing this extra middleman and buying wine straight from wineries is key to providing consumers lower prices, even when that means having to invest upfront in litigation or incur other costs to make a more direct, viable option. This strategy has been key to its ascent to become the top wine retailer in the entire country. Fighting to reduce the number of middlemen in a supply chain may take effort, but it can also result in meaningful savings for companies and their customers.

#3: ADD SOME DIRECT EXCHANGE TO THE MIX

A good starting point for exploring how to incorporate more direct exchange into your life is to examine ways it may already be part of your life and what it has meant to you. Maybe you have a local bakery that you visit every weekend for their delicious croissants. Or perhaps you have tried to buy food at a farmers' market only to watch it go to waste because you don't like cooking. There is no right answer here; the aim is to reflect on what you already know about yourself, and where direct exchange has and has not worked for you in the past.

The next step is to experiment. It can be helpful to be honest about strong preferences and aversions while also being open to stretching a bit. For example, after learning more about the chocolate industry, I decided to seek out a more ethical source to feed my cravings. I soon discovered Theo Chocolate, a Seattle-based chocolate company that issues an annual Impact Report with detailed information about where it sources its cocoa, the premium it pays—relative to both market and Fairtrade rates—and the lives of the farmers who grow that cocoa. It is a form of marketing, but one that is possible only because the company sourced all of the 1,500 metric tons of cocoa it used in 2020 from the community of Watalinga in the Democratic Republic of Congo. This type of focused sourcing allows it to build close working relationships with the farmers who grow the cocoa. Although transitioning to eating better chocolate is a luxury, getting comfortable spending more and making the effort to plan ahead has not come easily for me. Changing habits can be hard, even when there is a very sweet reward, so it can be helpful to have patience with the process.

Another great way to experiment with direct exchange is through travel. I found Navarro because I love visiting off-the-beaten-path wine destinations. Other people have other passions. Berea, Kentucky, former home of Churchill Weavers—the company that made acclaimed tapestries until it was acquired and shut down by Crown Crafts—remains a great place to buy an array of handcrafted goods directly from the artisans who make them. Located in the heart of Appalachia, it is home to hundreds of different workshops, many of which provide opportunities to see the process behind the crafts.

Conferences and fairs can provide a way to meet a lot of creators and get to know others with whom you share a common interest. For example, after funding the creation of his graphic novels on Kickstarter, Bob Frantz often sells them in-person at comic con events across the country. When such events span a few days or attract many of the same people year after year, they can also be great venues for building a deeper sense of community. Bob explained that it can often feel like the same twenty dollars just keeps moving in a "circle," as people buy comics off of each other online and at "cons"—yet another variation of community being created in a way that blends Hyde's depiction of how gift economies work with commerce.

The destination could also be closer to home. Independent breweries and small-scale coffee roasters have been proliferating in recent years. There were upwards of 8,000 small breweries in 2020, offering a place to gather with friends in addition to buying beer straight from the source. I learned quite a lot about the impact of roasting on coffee beans during a backyard lesson in my hometown of Ann Arbor, Michigan. The session was led by John Roos, a friend of a friend and founder of RoosRoast Coffee. I still buy a couple of bags of RoosRoast coffee at the farmers' market when I visit Michigan and it always makes me smile.

Reflect on your friends, family members, and other acquaintances. You may find a furniture maker, a self-published kids' book author, a sculptor, or some other creator who could use a little support and with whom you might enjoy deepening your connection. Personal ties can be a messy but meaningful way to experiment with direct exchange.

Do not expect that you will be able to go to the source in every aspect of your life. By mapping out the benefits and drawbacks of relying on middlemen, and how going to the source can sometimes unlock a whole different set of dynamics, this book also provides a road map that you can personalize to determine where you may want to experiment with change. You can use the insights from Part II regarding the benefits that middlemen provide to acknowledge the domains where, for now at least, those benefits are too great to forgo. The insights on the dark side of intermediation from Part III can help you identify instances when relying on middlemen has hidden costs, to you or others, that you no longer want to accept. And you can use the discussion in Part IV regarding the benefits unique to direct exchange to figure out where you may be able to unlock gift-like dynamics or find particular meaning in going to the source. This map will change over time, and it can take a while to figure out what belongs in what bucket for you, right now, but the very process of trying to map out these domains can be a useful way to put the different lessons from this book to work in your daily life.

Self-knowledge, a willingness to experiment, and mapping are also a good foundation for entrepreneurs and companies looking to incorporate direct into their business plans. It may be useful to start by identifying a particular aim, be it saving money, furthering sustainability goals, reducing supply chain risks, or something else entirely, and using that to identify the middlemen that potentially should be jettisoned in favor of a direct alternative. Use the map to assess the trade-offs in making a shift, and where the specific benefits of direct exchange—such as closer communication with suppliers or greater capacity to control the customer experience—are sufficient to justify that switch. If the risk seems too great, try using direct exchange as a complementary way of sourcing or selling goods or raising funds without shutting off established alternatives.

If there is no direct alternative where it seems like one could be viable, that may indicate opportunity. Even when direct options do exist, if they are priced or marketed only for a small slice of the population, that too may spell opportunity. For example, one reason I personally sometimes struggle to go direct or opt for the supply chain that is short enough to enable meaningful transparency is that the goods available via these routes are often luxurious in other ways as well. This is my struggle with Theo Chocolate, as much as I adore its chocolate. Although paying people a living wage and making things in smaller quantities can add significant costs, direct exchange should not be, and need not be, reserved for privileged few. Direct exchange for the many is sometimes an option, but not as often as it could or should be. Just as the rise of the middleman economy created opportunities, so too will the shift to a better equilibrium. Entrepreneurial efforts to find new ways to offer a good directly, fund a project directly, or, like Shopify, help others forge such connections are critical for ushering in a new era. The key is to look around not just at how things are, but with an eye to how they might be.

#4: FOLLOW THE FEES WHEN CHOOSING AMONG MIDDLEMEN

I have a confession: My first "real" job out of college was as a middleman. I was a stockbroker, under the haughtier title of Associate Financial Consultant, for the Lynnwood, Washington, office of Merrill Lynch. I was twenty-two and knew nothing about finance. Although I lasted less than a year, the experience taught me a lot about human nature and the power of salesmanship.

Neither Merrill Lynch nor my colleagues were looking to screw over clients. Merrill Lynch earned a lot of money from its "thundering herd" of brokers, and it knew that its reputation could suffer if its brokers put clients in blatantly bad investments. My colleagues had no ambition to become the next Wolf of Wall Street. Most just wanted to make their mortgage payments. And many of my colleagues provided good advice to clients on matters such as how much to save for retirement and the value of diversification.

But when it came to specifics of how to invest clients' money, subtle conflicts of interest came into play. There was no escaping that higher fees benefited Merrill Lynch, and Merrill compensated brokers accordingly. Clients, by contrast, often would have been best served by a less costly alternative. Accentuating the challenge, most people are more attuned to how their overall portfolio performs than slight differences in fee structures, so Merrill's concern with its reputation did little to constrain these subtle conflicts.

One way these subtle conflicts came into play was in the mutual funds that my colleagues would recommend to clients. Empirical research shows that most actively managed mutual funds do not earn a sufficient rate of return to justify their high fees, making them a bad choice for most investors.[11] Yet, precisely because actively managed funds charge so much, these funds generated meaningful fees that benefited Merrill Lynch and the brokers that sold them. Further greasing the wheels, the mutual fund companies charging the high fees would take us junior brokers out to lunch—or dinner or skiing—and explain why their product justified the high fees. They told us stories about what made their investment professionals so good and their approach so distinct. The stories we heard over those lunches made it easier for my colleagues to convince their clients— and themselves—that they were doing the right thing, despite the research to the contrary.

The point: Know how the middlemen in your life make their money. Know the biases they have and what alternatives exist. And recognize that sometimes paying a little more in advance can save you money and headaches down the road. Middlemen understand behavioral biases. They know that people often focus too much on up-front fees and too little on the ways they may pay indirectly or over time. Thinking ahead and asking the right questions can go a long way in knowing just how far, if at all, to trust a middleman's recommendations.

Organizations can put this principle to work by assessing where middlemen may be using small bribes to influence decision makers within their organization. Middlemen that charge overly high fees are often the ones best positioned to share the wealth with key decision makers. During the IPO heyday of the late 1990s, for example, Goldman Sachs and other investment banks allocated underpriced "hot" IPO shares to founders and executives at eBay, Yahoo, and other companies. When the price popped on opening day, the executives often sold the shares at a hefty profit.[12] This scheme made those individuals—the key decision makers who could influence what investment bank the company used in the future—more favorably inclined toward the investment banks that had provided them the shares, regardless of whether that bank was otherwise the best for their company. Evaluating who within an organization is getting wined, dined, or taken to free sporting events could reveal relationships with middlemen that are not optimal for your organization.

#5: HOW TO FIND AND BUILD BRIDGES

One way to use direct exchange to build bridges is to identify common challenges, shared passions, and other similarities that cut across racial and socioeconomic divides. Common sources of oppression can also be used as a foundation for building connections in ways that transcend place. Such approaches allow ingrained instincts to connect along a shared experience or interest to be used to bring about change, rather than affirm unjust allocations of power and wealth.

Many of the families with whom I have connected via GoFundMe have lives that look quite different from mine. They live across the country and around the world, and often in places far quieter than New York City. Yet in seeing our children through the various surgeries and other procedures they need to stay alive despite their unique hearts, we have something in common. I can still vividly recall the confidence in the eyes and stance of a little boy on the other side of the country who had outgrown earlier fixes to his heart and whose mother was seeking support for yet another surgery. Although physically he bore almost no resemblance to our daughter, I saw her in him. Our daughter too is outgrowing the makeshift valve in her heart; she too will need another surgery; and she too seems impervious to any of it. In the email exchange that followed the gift, his mother and I both promised to pray for each other's family. I still pray for her at times, along with her son and the many others who love him. We may live thousands of miles apart and have lives that differ in many and meaningful ways, but GoFundMe allowed money to flow where it was needed and it allowed us to connect in a way that still resonates.

These types of exchanges also bring home just how much luck contributes to our family's ability to access high-quality medical care and to take time away from work when our daughter needs us. They remind me of the need to keep fighting for a system where everyone has access to the care they need for themselves and those they love. Direct exchange can help get us outside the confines of our daily lives and enable us to see the society's structures and hierarchies from the perspective of someone who experiences them in a different way than we do. For some of us, some of the time, this is the perspective that we need to cultivate the humility and drive to keep advocating for change.

Some of the most inspiring examples of direct exchange bridge building come from people who themselves bridge multiple worlds. Hanahana Beauty, for example, deepens the roots that already connected Abena with Ghana and Chicago. Sana Javeri Kadri is helping to build her own transformative bridges with Diaspora Co., a spice company she founded. Diaspora sources spices directly from family farms, pays those farmers a higher price than they could get elsewhere, and sells those spices directly to consumers via a website that provides rich insights into the farmers, the spices, and who gets paid what. As someone who grew up in India, went to college and started working in the United States, and now splits her time between the two countries, Javeri Kadri was uniquely positioned to understand the needs of Indian spice farmers and what American consumers were missing out on when they settled for second-rate spices that lacked any sense of place. She used a Kickstarter campaign to help raise funds and soon started cultivating close relationships with farmers, helping them to transition toward more sustainable approaches to growing their spices, in addition to helping consumers appreciate those spices and creating a more equitable system of exchange. As Tamar Adler aptly summed up in a profile of Diaspora and Javeri Kadri: "Diaspora Co.'s single-origin spices . . . represent a new beginning, a reinvention of a spice route that rewards farmers directly for scrupulous farming, and rewards consumers with turmeric and black pepper (and more) that redefine the categories themselves."[13]

Shared passions are another great way for people to connect while moving outside their typical circles. Conferences and art fairs can offer this, but so too can makers. BrewDog, a beer start-up in Scotland, grew rapidly by building a community of devotees who drink its beer, provide money for it to grow, and give early product feedback. BrewDog also sponsors events, in person and virtual, where people gather to drink, enjoy life, and revel in their shared appreciation of BrewDog's brews. While this may not seem transformative in the same way as Hanahana, it too enables people to come together as people, and to develop a small piece of identity that lies apart from the professional, socioeconomic, and other statuses that so often operate to divide.

The examples here are a mere sampling of the ways that we can put the five principles to work in our daily lives and in our work. I am sure there are thousands of alternatives that I am not creative enough to imagine, and I look forward to watching those alternatives also spring to life. Yet individual action and entrepreneurial vision alone are not enough to bring about the changes we need. The threats middlemen pose are too great and too pervasive to be tackled without help from policy makers. Fortunately, the same five principles can be used to identify the types of policies needed to contain the dangers posed by the middleman economy and lay the foundation for a more direct, resilient, and accountable system.

POLICIES THAT PROTECT PEOPLE, NOT MIDDLEMEN

The middleman economy isn't just an economy that uses middlemen to play certain roles; it is one defined by overly large, overly powerful middlemen and long, complex chains. It is an economy where the most frequently trod paths involve *both* large middlemen and complex supply chains. It is an economy where people are seemingly saturated with information, and yet often don't know—and have no way to learn—the first thing about the actual people and places behind the goods they are consuming, or the people, places, and activities that they are helping to fund when they invest their retirement savings. It is an economy that provides tangible benefits, including cheap goods, seemingly cheap ways to trade and invest, and mind-blowing convenience, but those benefits often come with hidden costs for the user, other people, and the planet. It is an economy laden with hidden frailties that only become evident during periods of distress, just when we most need our systems of production and capital formation to work well. It is an economy where workers increasingly work for middlemen or in other roles that are so specialized as to blind workers, too, from the way their efforts may help to serve others.

The middleman economy was never inevitable and policy makers deserve some of the blame for allowing it to grow and morph as it has. For far too long, policy makers have tended to protect and promote this system. Sometimes this was the result of self-interest; other times, it was a by-product of the wealth of information and resources that middlemen bring to the table. Changing these dynamics is not easy, but it is possible. What follows is an exemplary, rather than exhaustive, overview of ways that policy makers can use the five principles to craft policies that help the economy better serve real people.

#1: WHY INTERMEDIATION SHOULD MATTER TO POLICY MAKERS

There are laws specifically designed to counter excessive and dangerous concentrations of power. They are the antitrust laws already touched on at various points throughout the discussion. Antitrust allows regulators to block anticompetitive mergers, limit how dominant actors can wield their power, and prohibit coordination meant to suppress competition. One way for policy makers to put "intermediation matters" into action is to ensure that existing antitrust laws are applied with rigor to the giant middlemen and networks of middlemen that sit at the core of the middleman economy.

There has been some meaningful recent progress in this direction. There are fresh efforts to use antitrust laws to bring about further reform in the real estate market, for example. And in 2019, a congressional subcommittee undertook a major, bipartisan investigation into the four tech giants. Although only Amazon is a middleman as defined here, at least with respect to its core business, all of the big four tech companies function at times like middlemen and can possess power and pose threats for reasons akin to the dynamics at play with large middleman. One of the most hopeful developments was President Joe Biden's decision to appoint Lina Khan, a vocal advocate for more robust antitrust enforcement who gained early fame for an article explaining why Amazon poses a serious threat that antitrust ought to take seriously, to chair the FTC, the leading antitrust regulator.[14]

The analysis here provides fresh support for efforts to use antitrust to tackle the threats posed by Amazon, real estate agents, and others, and it provides additional insight into the types of remedies needed to bring about real change. The disparity between the transformation in real estate brokerage that the Justice Department expected from its 2008 consent decree and the far more modest improvements that were actually realized is instructive. The advantages that large middlemen and networks of middlemen enjoy are so significant that they cannot easily be displaced or put into place. Real transformation is possible, but it requires antitrust authorities to have a deep understanding of how a particular middleman operates, an appreciation of the many sources from which it derives power, and the courage and creativity to tackle those sources at their core.

For example, with respect to Amazon, it may be necessary to find a way to disentangle Amazon-the-seller from Amazon-the-platform-for-other-sellers. One way to do this might be to allow both to continue to use the same distribution system, while forcing each to create a separate consumer-facing website and restricting their ability to share data or otherwise coordinate. In short, understanding the many sources of Amazon's influence may reveal that only Amazon can provide a check on the power now held by Amazon.

Antitrust is far from the only tool that could be used more aggressively and strategically to address the challenges posed by today's middleman economy, and how best to use this tool remains an open question. Nonetheless, the progress and potential in this domain illustrate how understanding middlemen and intermediation could help policy makers identify better ways to use the tools they already possess to promote the well-being of ordinary Americans.

#2: POLICIES TO PROMOTE SHORTER CHAINS

There are numerous policies that could help facilitate a shift toward the shorter supply chains needed to enhance resilience and accountability. For example, disclosure rules that require makers to provide information about environmental impact or the conditions of laborers could help—if they are designed with the aim of bringing about shorter chains and changes in where and how things are made. As reflected in the unintended consequences that flowed from the conflict minerals rule and the many other disclosure regimes that fail to live up to their aims, merely layering disclosure obligations on top of a long supply chain is not the answer. Instead, the aim must be to use disclosure as one tool among many with the express aim of shortening chains. For example, disclosure obligations could be one element of a regulatory strategy that imposes less onerous disclosure requirements for producers that use shorter, more transparent supply chains.

Policy makers can also craft other regulatory burdens to discourage overly long supply chains. There are some domains where this is happening already. For example, after CDOs backed by MBSs performed so poorly during the 2008 financial crisis, bank regulators imposed higher capital requirements on securitized assets backed by securitized assets.[15] Imposing this type of rule earlier—which may have seemed wise even before 2008, had regulators been focused on the length and complexity of the capital supply chain—may well have reduced the number of CDOs ever created and the fragility they helped perpetuate. Far more could be done along these lines.

Additional layers in the chains through which capital flows can increase the rigidities and information gaps that undermine resilience. That additional nodes are often the by-product of efforts by middlemen to increase their pay or minimize regulatory burdens suggests regulators could often discourage additional nodes with minimal adverse consequence. Financial regulators could be well served to map the layers of middlemen through which money flows, identify the supposed value proposition for each, and then systematically seek to reduce the number of layers, starting with those most likely to contribute to fragility and consumer harms.

Yet another way policy makers could promote shorter supply chains and more direct exchange is by helping to provide public or otherwise subsidized infrastructure to facilitate the movement of goods and capital. For example, small-scale sellers that don't want to give control over to Amazon must rely on third parties to ship goods to their customers. When the U.S. Postal Service failed to deliver a lot of packages in time for Christmas 2020, small sellers on Etsy were among those harmed. Ensuring that the USPS remains a robust option, and perhaps even providing subsidies for small businesses when they use USPS to send goods to customers, could help level the playing field.

#3: POLICIES THAT SUPPORT DIRECT EXCHANGE

Supporting the USPS is just one example of how the government can help provide the infrastructure needed to facilitate direct exchange. Other efforts in this vein can be even more targeted, such as creating subsidized platforms that facilitate direct exchange. Many towns already support farmers' markets and street fairs where makers can sell their goods directly by enabling the use of public spaces. This is a great start and a model that could be expanded and built upon.

Policy makers can also create new types of infrastructure. For example, one movement that has yet to gain widespread traction is "locavesting"—investing local.[16] The original idea grows out of the shop local movement: Just as many people are content to pay a little more for a good in order to support local businesses, there are probably people who would accept a lower rate of return or more risk in order to provide local businesses the capital they need to survive and thrive. This type of direct investment opportunity could also cultivate community, helping local businesses and the residents supporting them feel more connected. Yet the logistical challenges, from trying to structure the investments in accord with federal securities laws to the logistics of aggregating and moving the moneys involved, have so far prevented locavesting from taking off like "shop local." These are the types of functions often played by middlemen, and private innovation may one day offer easy ways to overcome these challenges. Nonetheless, because of the many public and nonpecuniary benefits that flow from this type of direct investment, private innovation is likely to be far below the optimal level. This creates an opportunity for enterprising state and local governments to try to build the pipes and other infrastructure needed to facilitate local investment. These types of initiatives are not easy to launch, but the range of potential benefits shows why they may well be worth the effort, particularly if different jurisdictions can find ways to work together and learn from one another.

Another way that policy makers can facilitate direct exchange is by removing barriers. Very often, people selling goods or seeking investors are subject to licensing or other requirements. Although well intentioned and often quite helpful, the burden of such requirements can be a major hurdle for smaller creators and entrepreneurs. Creating *de minimis* exceptions that exempt smaller actors from such obligations can enable more small-scale production and activity, where direct is often most powerful.

Policy makers should also be on the lookout for other ways that the law currently favors middlemen. Donations are a prime example. When someone gives money to a charity, that donation is tax deductible, even though much of the money given goes to administrative costs and further fund-raising. By contrast, when someone gives directly to a stranger on GoFundMe, none of the money given is tax deductible. There are some exceptions, and some good reasons not to make it too easy for people to get tax deductions, particularly when they could game the system. Yet this is another area where a small-scale exception could encourage people to experiment with providing direct support without enabling too much gamesmanship.

#4: FOLLOWING THE FEES WHEN CRAFTING POLICY

Policy makers have two important roles to play in putting this principle into practice: helping consumers to follow the fees and following the fees themselves. With respect to helping consumers and investors, middlemen often have no obligation to disclose how they are compensated. Even when fee structures are disclosed, consumers typically don't have any way of knowing how those fees compare with the fees a middleman would have earned on an alternative product. For example, a financial adviser selling a mutual fund has to disclose the fees charged by the mutual fund, but he doesn't have to tell his clients about index funds that would have cost them far less, and earned him less accordingly. Similarly, when shopping for a mattress, a shopper has no way of knowing whether the salesman really thinks a particular mattress is good or if he is just saying that because he earns the biggest commission on that brand. This puts consumers and investors at a real disadvantage and makes it easier for middlemen to abuse the trust placed in them.

Well-designed disclosure rules could help. Providing information about how middlemen are compensated is a start. Further requiring that middlemen provide information about the fees associated with an alternative "baseline" product could be even more effective in empowering consumers, investors, and other decision makers. This additional type of disclosure requirement may be warranted when the decision is big enough or the nature of the decision environment makes it ripe for abuse.

Another implication of "follow the fees" is that policy makers should at times limit the types of fee structures that middlemen can use. For example, neutrality principles have often been used to limit the ability of actors that provide critical services to discriminate among users. Their origins go back to common carriers, but they have also been used more recently to limit the ability of Internet service providers, for example, to discriminate among different content providers. Applying this doctrine to today's middlemen could entail limits on the ability of a dominant middleman to deny access to its platform, limiting a dominant platform's use of certain types of fee structures, or limiting the factors a dominant platform can incorporate into the algorithms that determine which sellers actually get seen by interested consumers. It could also be used to provide access to other infrastructure, such as the MLS, Amazon Marketplace, or a distribution network, so long as reasonable compensation is provided.

Although middlemen need to earn something on infrastructure in order to be willing to build it in the first place, there is little to suggest that the rewards should be as rich or as persistent as they are today.

And, as the history of neutrality principles shows, sometimes limiting the capacity of a critical access point to discriminate can promote flourishing and innovation at other points in a chain.

There may also be areas where policy makers should blend these approaches. Real estate is a great example of a domain where the normal fee structure could well be perpetuating a suboptimal outcome. As we saw, because sellers effectively pay the buyer's agent, and the buyer's agent has a lot of influence over which houses his clients see, it is very hard for individual sellers to offer buyers' agents anything less than the going rate in the area, regardless of what value they add. Bringing transparency to these price structures—as Trelora tried to do—would be a helpful first step, but more may well be required. For example, state lawmakers could impose a limit on how much sellers can pay to a buyer's agent—say, no more than $5,000 or 1 percent of the value of the home sold. This would not be a hard cap, as buyers' agents would remain free to charge their own clients any amount those clients are willing to pay for their services. Nonetheless, by reducing the need for sellers to overpay buyers' agents just to avoid having their home discriminated against and forcing buyers to actually experience the fees that they are effectively paying, such an intervention could go a long way in helping to ensure that the fees paid better reflect the value of the services provided.

Policy makers also need to understand how middlemen make money in order to know how that influences their lobbying and other efforts to shape lawmaking. As the chapters exploring the dark side of the middleman economy revealed, the law too often protects middlemen and outdated modes of intermediation. This is in part because middlemen really do understand the relevant markets and can provide useful insights into unintended consequences. To be effective, policy makers need more than backbone and a generic understanding that all lobbyists have biases; they need to understand how a particular middleman makes its money and how a potential intervention would affect its business model. This is key to enabling lawmakers to better understand when and how much to discount the tall tales that middlemen and their lobbyists can sometimes spin to protect a scheme from which they benefit. Less targeted efforts to limit lobbying and campaign contributions would also be helpful in reducing the advantages that middlemen too often enjoy.

#5: HOW POLICY MAKERS CAN HELP BUILD BRIDGES

Many of the ways that policy makers could work to promote direct exchange could be modified to promote the creation of new bridges. The government could, for example, subsidize the creation of online platforms or in-person fairs specially designed to promote small-production makers of color or to increase awareness around a particular group. The government could also build on existing programs that already allow federal food support to be used at farmers' markets, expanding the types of consumers that can readily access food direct from farmers.

States and municipalities could also set up trial programs and events based on feedback from constituents looking to use direct exchange in transformative ways. They could start by holding events where they learn from makers, creators, and entrepreneurs in their area about the challenges they face, and what type of support they need to forge direct connections. Part of building bridges means listening to those who too often have not had the microphone and letting them play a role setting the agenda.

Separately, policy makers can use their authority to reduce the capacity of middlemen to perpetuate division and inequity. For example, there are many laws on the books that are meant to preclude real estate agents and those engaged in housing finance from engaging in discriminatory behavior of the kind that so often leaves neighborhoods segregated and otherwise limits the ability of Black and Hispanic families to use home ownership to build wealth at the same rate as white families. Yet enforcement has often been limited and challenging. Both better laws and more robust enforcement of existing laws could help reduce the harms that flow from the middleman economy, and could indirectly bolster better, more direct alternatives.

The middleman economy cannot be blamed for all of the ills facing society today, but it contributes to many of them. By seeing the middleman economy for what it is and understanding how middlemen have wielded their power to suppress competition, bias consumer decision making, and contort lawmaking, lawmakers can position themselves to help chart a new course.

THE MOST IMPORTANT lesson for anyone concerned about the middleman economy is to do something. The problem is so big that it can be easy to feel overwhelmed. But each choice matters. Understanding the middleman economy is the first step in figuring out where we can make different, and better, choices. Adding direct to the mix shows that opting to bypass middlemen is not just a chore undertaken for the greater good, but an opportunity to forge new connections, live out values, and possibly discover some unexpected joys.

If there are areas of your life where change feels too hard, that too can be a signal for action, but of a different sort. Many of the challenges created by today's middleman economy are collectively borne, even if we each experience them in our own, individualized way. One reason I wrote this book is that I found it just too hard, much of the time, to avoid giant middlemen despite my growing discomfort in using them. Giving public voice to these concerns can help sow the seeds needed to bring about systemic change.

If you are feeling bold, consider writing a letter—or perhaps an email—to your elected officials. From limiting the scale and scope of the largest middlemen to subsidizing the development of direct alternatives and bringing about the structural changes needed to enable real transparency, there are a lot of tools that the government alone can wield. The best way to get Congress, state legislatures, and other policy makers to take action is to let them know that you care.

Sharing frustrations and success stories with others, and learning from their experiences, can also be helpful. If you have a direct producer that you think is great, tell your friends. Or post it on social media. I would love to see the various ways that you go to the source, whether it is a new experience or something you have been doing for a long time but now see in a new light. Talking to friends and hearing about the experiences of others can also be key to getting through rough times. Given the scale of today's middleman economy, opting out is not easy, no matter how much we may want to. Although my life feels richer as a result of incorporating the five principles into my decision making, I still struggle, sometimes daily, to know what is right and to live accordingly.

Changing habits is hard. Changing economic structures is harder still. The good news is that none of us is in this alone. The stories told here are but a tiny slice of the myriad, beautiful ways that people are already going to the source. Small changes can add up to big results. Each time we buy, invest, or give directly, each time we seek out or help to build a shorter supply chain, each time we support one another in our collective effort to do things differently, and each time we advocate for reform, we are helping to dismantle the middleman economy and replace it with a system that is more just, more resilient, and more humane.

| ACKNOWLEDGMENTS |

A BOVE ALL, I want to thank my husband, Tim Wu, whose love, support, faith, and feedback helped make this book possible. Coming home to you, Sierra, and Essie is just the elixir I need to return to work with a fresh eye. I also want to thank all of the Judges, Wolfs, Sellmyers, Wus, and other dear family members who have provided us so much support as I worked on this project. Family has always gotten me through the good times and the bad, and there is nothing like a pandemic to make me particularly grateful for mine.

I am indebted to my agent, Laurie Abkemeier, for providing valuable feedback and guidance at every stage of this project. I am also grateful to Hollis Heimbouch, Rebecca Raskin, Kirby Sandmeyer, and the entire team at HarperBusiness for sharing my vision for this book and helping to make it a reality.

This book brings together years of research that benefited from conversations and feedback from more colleagues and friends than I can possibly name here. I particularly want to thank Jedediah Britton-Purdy, Barbara Burton, Tom Morton, Charles Sabel, and Rory Van Loo, whose comments on earlier drafts were pivotal in helping me understand the project, including its shortcomings. I also want to thank the many wonderful research assistants at Columbia Law School who helped on various stages of this project and the research feeding into it, including Skanda Amarnath, Oluwatumise Asebiomo, Clare Curran, Connor Clerkin, Eddie Kim, Johannes Liefke, Alex Perry, Jordan Schiff, Ethan Stern, and Jake Todd.

| NOTES |

PREFACE: THE QUIET TRANSFORMATION

1. "Fortune 500 2021," *Fortune*, https://fortune.com/fortune500/, accessed Oct. 27, 2021.

2. Taylor Sopor, "Amazon now employs nearly 1.3 million people worldwide after adding 500,000 workers in 2020," *GeekWire*, Feb 22, 2021, https://www.geekwire.com/2021/amazon-now-employs-nearly-1-3-million-people-worldwide-adding-500000-workers-2020/; "How Many People Work at Walmart?," *Walmart Corporate - Ask Walmart*, as last modified March 1, 2021, https://corporate.walmart.com/askwalmart/how-many-people-work-at-walmart.

3. Krystina Gustafson, "Nearly Every American Spent Money at Wal-Mart Last Year," *CNBC*, April 12, 2017, https://www.cnbc.com/2017/04/12/nearly-every-american-spent-money-at-wal-mart-last-year.html.

4. Lucy Handley, "Amazon's brand value tops $400 billion, boosted by the coronavirus pandemic: Survey," *CNBC*, June 30, 2020, https://www.cnbc.com/2020/06/30/amazons-brand-value-tops-400-billion-according-to-kantar-report.html; Kantar BrandZ Most Valuable Global Brands Report 2021, https://www.kantar.com/campaigns/brandz/global.

5. Giacomo Tognini, "After Two Weeks At No. 2, Jeff Bezos Is Once Again the Richest Person in the World," *Forbes*, June 10, 2021, https://www.forbes.com/sites/giacomotognini/2021/06/10/after-two-weeks-at-no-2-jeff-bezos-is-once-again-the-richest-person-in-the-world/?sh=34d6c1a757aa.

6. Tom Metcalf, "These Are the World's Richest Families," *Bloomberg*, August 1, 2020, https://www.bloomberg.com/features/richest-families-in-the-world/?sref=0SF97H1m.

7. This saying is often attributed to Albert Einstein, but I could find no source able to trace its true genesis. I take the attribution story to reflect the wisdom people see in the insight, which may be as important as its true origins.

CHAPTER 1: THE HIDDEN COST OF CONVENIENCE

1. Jeffrey T. McCollum et al., "Multistate Outbreak of Listeriosis Associated with Cantaloupe," *New England Journal of Medicine* 369 (2013): 944–53.

2. C. Frank et al., "Large and Ongoing Outbreak of Haemolytic Uraemic Syndrome, Germany, May 2011," *Euro Surveillance* 16 (2011), https://edoc.rki.de/bitstream/handle/176904/882/23biStyp7ZDrU.pdf?sequence=1&isAllowed=y.

3. Kai Kupferschmidt, "Cucumbers May Be Culprit in Massive *E. coli* Outbreak in Germany," *Science*, May 26, 2011.

4. "Ehec–Woher stammt Erreger O104?," DIE ZEIT, June 16, 2011; "Warnungen wegen Ehec–'Ich kaufe und esse alles,'" SÜDDEUTSCHE ZEITUNG, June 10, 2011. Translated for author.

5. "Einnahmeausfall–Handel fordert EHEC-Entschädigung," MANAGER MAGAZIN, June 19, 2011. ("Ich werde schon wie ein potentieller Mörder behandelt, nur weil ich Gurken und Tomaten verkaufe.")

6. Ibid.

7. Karch et al., "The Enemy Within Us: Lessons from the 2011 European *Escherichia coli* O104:H4 Outbreak," *EMBO Molecular Medicine* 4 (2012): 841–48, http://embomolmed.embopress.org/content/embomm/4/9/841.full.pdf.

8. "Dutch Join GB in Hamburg Rowing World Cup Pull-Out," *BBC Sport*, BBC, June 9, 2011, https://www.bbc.com/sport/rowing/13718566.

9. Udo Buchholz et al., "German Outbreak of *Escherichia coli* O104:H4 Associated with Sprouts," *New England Journal of Medicine* 365 (2011): 1763–70.

10. Helge Karch et al., "The Enemy Within Us," 841.

11. Ibid.

12. Michael Moss, "The Burger That Shattered Her Life," *New York Times*, October 3, 2009, https://www.nytimes.com/2009/10/04/health/04meat.html; "Real Life Impacts of *E. coli* infection and HUS," Marler Clark LLP, https://about-ecoli.com/real_life_impacts.

13. Moss, "The Burger That Shattered Her Life."

14. Michael Moss, *Salt, Sugar, Fat: How the Food Giants Hooked Us* (New York: Random House, 2013), xxiv.

15. Fabrizio Dabbene, Paolo Gay, and Cristina Tortia, "Traceability Issues in Food Supply Chain Management: A Review," *Biosystems Engineering* 120 (2014): 65–80.

16. *Empty Promises: The Failure of Voluntary Corporate Social Responsibility Initiatives to Improve Farmer Incomes in the Ivorian Cocoa Sector*, Corporate Accountability Lab (July 2019), https://static1.squarespace.com/static/5810d-da3e3df28ce37b58357/t/5d31c76d06b158000167f385/1563543422666/Empty_Promises_2019pdf.pdf; *The Cocoa Protocol: Success or Failure?*, International Labor Rights Forum (June 30, 2008), https://laborrights.org/sites/default/files/publications-and-resources/Cocoa%20Protocol%20Success%20or%20Failure%20June%202008.pdf.

17. Chocolate Manufacturers Association, *Protocol for the Growing and Processing of Cocoa Beans and Their Derivative Products in a Manner That Complies with ILO Convention 182 Concerning the Prohibition and Immediate Action for the Elimination of the Worst Forms of Child Labor*, International Cocoa

Initiative (December 8, 2015), https://web.archive.org/web/20151208022828
/http://www.cocoainitiative.org/en/documents-manager/english/54-harkin
-engel-protocol/file.

18. Anthony Myers, "New Report Reveals Child Labor on West African Cocoa
    Farms Has Increased in Past 10 Years," *Confectionery: Sustainability* (May 7,
    2020), https://www.confectionerynews.com/Article/2020/05/07/New-report
    -reveals-child-labor-on-West-African-cocoa-farms has increased-in-past-10
    -years; J. Edward Moreno, "US Report on West African Child Labor Facing
    Review Following Objections," *Hill*, June 12, 2020, https://thehill.com/policy
    /international/africa/502444-us-report-on-west-african-child-labor-facing
    -review-following.

19. *Final Report 2013–14: Survey Research on Child Labor in West African Co-
    coa Growing Areas*, Tulane University School of Public Health, Payson Cen-
    ter for International Development (July 30, 2015), https://www.dol.gov/sites
    /dolgov/files/ILAB/research_file_attachment/Tulane%20University%20
    -%20Survey%20Research%20Cocoa%20Sector%20-%2030%20July%20
    2015.pdf.

20. *Empty Promises*, Corporate Accountability Lab.

21. Peter Whoriskey and Rachel Siegel, "Cocoa's Child Laborers," *Washington
    Post*, June 5, 2019, https://www.washingtonpost.com/graphics/2019/business
    /hershey-nestle-mars-chocolate-child-labor-west-africa/.

22. "World's Largest Chocolate Companies Rated on Efforts to End Environ-
    mental and Labor Abuses," *Green America: Labor*, April 7, 2020, https://
    www.greenamerica.org/press-release/chocolate-companies-rated-addressing
    -environmental-labor-abuses.

23. Ibid.

24. "Child Labor in Your Chocolate? Check Our Chocolate Scorecard," Green
    America, October 16, 2019, https://www.greenamerica.org/end-child-labor-
    cocoa/chocolate-scorecard#fn1.

25. *Empty Promises*, Corporate Accountability Lab.

26. *Hodsdon v. Mars, Inc.*, 891 F.3d 857 (9th Cir. 2018), http://cdn.ca9.uscourts.
    gov/datastore/opinions/2018/06/04/16-15444.pdf; *Laura Dana v. The Her-
    shey Company*, 730 F. App'x 460 (9th Cir. 2018), https://cdn.ca9.uscourts.gov
    /datastore/memoranda/2018/07/10/16-15789.pdf; *Elaine McCoy v. Nestlé
    USA, Inc.*, No. 16–15794 (9th Cir. 2018), https://cdn.ca9.uscourts.gov/data
    store/memoranda/2018/07/10/16-15794.pdf.

27. *Hodsdon v. Mars, Inc.*, 4.

28. Derek Thompson, "How America Spends Money: 100 Years in the Life of the
    Family Budget," *Atlantic*, April 5, 2012, https://www.theatlantic.com/business
    /archive/2012/04/how-america-spends-money-100-years-in-the-life-of-the
    -family-budget/255475/.

29. Ibid.

30. "The Number of U.S. Farms Continues to Decline Slowly," U.S. Depart-
    ment of Agriculture, Economic Research Service, last modified May 10, 2021,
    https://www.ers.usda.gov/data-products/chart-gallery/gallery/chart-detail
    /?chartId=58268.

31. "Farming and Farm Income," U.S Department of Agriculture, Economic Re-
    search Service, last modified September 2, 2021, https://www.ers.usda.gov

/data-products/ag-and-food-statistics-charting-the-essentials/farming-and
-farm-income/.

32. Ibid.

33. Marco Margaritoff, "Drones in Agriculture: How UAVs Make Farming More Efficient," The Drive, February 13, 2018, http://www.thedrive.com/tech/18456 /drones-in-agriculture-how-uavs-make-farming-more-efficient.

34. PwC Poland, *Clarity from Above: PwC Global Report on the Commercial Applications of Drone Technology* (May 2016), 4, https://www.pwc.pl/pl/pdf /clarity-from-above-pwc.pdf; Food and Agriculture Organization of the United Nations and International Telecommunication Union, *E-Agriculture in Action: Drones for Agriculture* (2018), 27, http://www.fao.org/3/I8494EN /i8494en.pdf.

35. "Farming and Farm Income," USDA.

36. "Agricultural Trade," U.S. Department of Agriculture, Economic Research Service, last modified August 20, 2019, https://www.ers.usda.gov/data-products /ag-and-food-statistics-charting-the-essentials/agricultural-trade/.

37. "Percentage of U.S. Agricultural Products Exported," U.S. Department of Agriculture, Foreign Agricultural Service, May 30, 2018, https://www.fas.usda .gov/data/percentage-us-agricultural-products-exported.

38. Vijaya Chebolu-Subramaniana and Gary M. Gaukler, "Product Contamination in a Multi-Stage Food Supply Chain," *European Journal of Operational Research* 244 (2015): 164–75.

39. Kelly Egolf, "Locavore," *Verde*, October 15, 2014, http://verdefood.com /locavore/.

40. "Assets, Debt and Wealth," U.S Department of Agriculture, Economic Research Service, last modified September 2, 2021, https://www.ers.usda.gov /topics/farm-economy/farm-sector-income-finances/assets-debt-and -wealth/.

41. "Farm Bankruptcies Rise Again," American Farm Bureau Federation, October 30, 2019, https://www.fb.org/market-intel/farm-bankruptcies-rise-again; Jesse Newman, "More Farmers Declare Bankruptcy Despite Record Levels of Federal Aid," *Wall Street Journal*, August 6, 2020, https://www.wsj.com /articles/more-farmers-declare-bankruptcy-despite-record-levels-of-federal -aid-11596706201.

42. Joe Wertz, "Farming's Growing Problem," Center for Public Integrity, January 22, 2020, https://publicintegrity.org/environment/unintended-consequences -farming-fertilizer-climate-health-water-nitrogen/; Peiyu Cao, Chaoqun Lu, and Zhen Yu, "Historical Nitrogen Fertilizer Use in Agricultural Ecosystems of the Contiguous United States During 1850–2015: Application Rate, Timing, and Fertilizer Types," *Earth System Science Data* 10, no. 2 (June 4, 2018): 969–84, https://doi.org/10.5194/essd-10-969-2018.

43. Jason Hill et al., "Air-Quality-Related Health Damages of Maize," *Nature Sustainability* 2 (April 1, 2019): 297–403, https://www.nature.com/articles/s41893 –019–0261-y.

44. Tariq Khokhar, "Chart: Globally, 70% of Freshwater Is Used for Agriculture," World Bank Blogs, March 22, 2017, https://blogs.worldbank.org/opendata /chart-globally-70-freshwater-used-agriculture.

45. J. Poore and T. Nemecek, "Reducing Food's Environmental Impacts Through Producers and Consumers," *Science* 360, no. 6392 (June 1, 2018): 987–92, https://science.sciencemag.org/content/360/6392/987.

46. Jonathan Watts, "Third of Earth's Soil Is Acutely Degraded Due to Agriculture," *Guardian*, September 12, 2017, https://www.theguardian.com/environment/2017/sep/12/third-of-earths-soil-acutely-degraded-due-to-agriculture-study.

47. U.S. Department of Agriculture, Economic Research Service, *America's Diverse Family Farms: Economic Information Bulletin No. 203* (December 2018): 5–6, https://www.ers.usda.gov/webdocs/publications/90985/eib-203.pdf?v=9520.4.

48. Ibid.

49. Jules Scully, "The 2019 Top 100 Food & Beverage Companies," *Food Engineering*, September 9, 2019, https://www.foodengineeringmag.com/articles/98481-the-2019-top-100-food-beverage-companies.

50. "Our Story," Blue Bottle Coffee, https://bluebottlecoffee.com/our-story, accessed June 21, 2021.

51. Roland Schroll, Benedikt Schnurr, and Dhruv Grewal, "Humanizing Products with Handwritten Typefaces," *Journal of Consumer Research* 45 (2018), doi:10.1093/jcr/ucy014 at 649.

52. Cargill, *Cargill 2018 Annual Report* (2018), 4, https://www.cargill.com/doc/1432124831909/2018-annual-report.pdf.

53. Ibid., 5.

54. Lawrence Lessig, *Republic, Lost: Version 2.0* (New York: Grand Central, 2015).

55. "Agribusiness: Long-Term Contribution Trends," OpenSecrets.org: The Center for Responsive Politics, accessed March 17, 2021, https://www.opensecrets.org/industries/totals.php?cycle=2020&ind=A.

56. FedByTrade, https://www.cargill.com/fedbytrade, accessed October 27, 2021.

57. Marion Nestle, *Food Politics: How the Food Industry Influences Nutrition and Health* (Berkeley: University of California Press, 2013).

58. "Agribusiness Is the Biggest Lobbyist on the EU-US Trade Deal, New Research Reveals," Corporate Europe Observatory, August 7, 2014, https://corporateeurope.org/pressreleases/2014/07/agribusiness-biggest-lobbyist-eu-us-trade-deal-new-research-reveals.

59. "FAQ on the Pew Commission on Industrial Farm Animal Production," Pew Charitable Trusts, October 23, 2013, https://www.pewtrusts.org/en/research-and-analysis/articles/2013/10/22/faq-on-the-pew-commission-on-industrial-farm-animal-production.

60. Pew Commission on Industrial Farm Animal Production, *Putting Meat on the Table: Industrial Farm Animal Production in America* (2008), 3, http://www.pcifapia.org/_images/PCIFAPFin.pdf.

61. Ibid., 5.

62. Ibid., 6.

63. "FAQ on the Pew Commission on Industrial Farm Animal Production," Pew Charitable Trusts.

64. Pew Commission on Industrial Farm Animal Production, *Putting Meat on the Table*, viii.

65. Ibid.

66. Johns Hopkins Center for a Livable Future, *Industrial Food Animal Production in America: Examining the Impact of the Pew Commission's Priority Recommendations* (2013), https://clf.jhsph.edu/sites/default/files/2019-05/industrial-food-animal-productionin-america.pdf.

67. Ibid., 46.

68. Helena Bottemiller Evich, "Meat Industry Wins Round in War Over Federal Nutrition Advice," *Politico*, January 7, 2016, https://www.politico.com/story/2016/01/2015-dietary-guidelines-217438#.wthnpnn:IETp; Markham Heid, "Experts Say Lobbying Skewed the U.S. Dietary Guidelines," *Time*, January 8, 2016, http://time.com/4130043/lobbying-politics-dietary-guidelines/. See also U.S. Department of Agriculture, Agriculture Research Service, *Scientific Report of the 2015 Dietary Guidelines Advisory Committee: Advisory Report to the Secretary of Health and Human Services and the Secretary of Agriculture* (2015), https://health.gov/dietaryguidelines/2015-scientific-report/pdfs/scientific-report-of-the-2015-dietary-guidelines-advisory-committee.pdf.

69. Zephyr Teachout, *Break 'Em Up: Recovering Our Freedom from Big Ag, Big Tech, and Big Money* (New York: All Points Books, 2020), 19.

70. Polly Mosendz, Peter Waldman, and Lydia Mulvany, "U.S. Meat Plants Are Deadly as Ever, with No Incentive to Change", *Bloomberg Law*, June 18, 2020, https://news.bloomberglaw.com/daily-labor-report/u-s-meat-plants-are-deadly-as-ever-with-no-incentive-to-change.

71. Kimberly Kindy, "More Than 200 Meat Plant Workers in the U.S. Have Died of COVID-19. Federal Regulators Just Issued Two Modest Fines," *Washington Post*, September 13, 2020, https://www.washingtonpost.com/national/osha-covid-meat-plant-fines/2020/09/13/1dca3e14-f395-11ea-bc45-e5d48ab44b9f_story.html.

72. "U.S. Department of Labor Cites Smithfield Packaged Meats Corp. for Failing to Protect Employees from Coronavirus," Occupational Safety and Health Administration, U.S. Department of Labor, September 10, 2020, https://www.osha.gov/news/newsreleases/region8/09102020; "U.S. Department of Labor Cites JBS Foods Inc. for Failing to Protect Employees from Exposure to the Coronavirus," Occupational Safety and Health Administration, U.S. Department of Labor, September 11, 2020, https://www.osha.gov/news/newsreleases/region8/09112020.

73. Dave Mead et al., "The Impact of the COVID-19 Pandemic on Food Price Indexes and Data Collection," *Monthly Labor Review*, U.S. Bureau of Labor Statistics, August 2020, https://doi.org/10.21916/mlr.2020.18.

74. Rakesh Kochhar, "Unemployment Rose Higher in Three Months of COVID-19 Than It Did in Two Years of the Great Recession," Factank: News in the Numbers, Pew Research Center, June 11, 2020, https://www.pewresearch.org/fact-tank/2020/06/11/unemployment-rose-higher-in-three-months-of-covid-19-than-it-did-in-two-years-of-the-great-recession/.

75. Sophie Kevany, "Millions of US Farm Animals to Be Culled by Suffocation, Drowning and Shooting," *Guardian*, May 19, 2020, https://www.theguardian.com/environment/2020/may/19/millions-of-us-farm-animals-to-be-culled-by-suffocation-drowning-and-shooting-coronavirus.

76. Richard Hall, "2018 'Another Record Year' for Food and Beverage Acqui-

sitions," *FoodBev Media*, January 17, 2019, https://web.archive.org/web
/20190117135916/https://www.foodbev.com/news/2018-another-record-year
-for-food-and-beverage-acquisitions/.

CHAPTER 2: THE JOY OF GOING TO THE SOURCE

1. Craig J. Thompson and Gokcen Coskuner-Balli, "Enchanting Ethical Consumerism: The Case of Community Supported Agriculture," *Journal of Consumer Culture* 7, no. 3 (2007): 275–303.

2. U.S. Department of Agriculture, National Agricultural Library, Alternative Farming Systems Information Center, *Community Supported Agriculture*, last reviewed September 2021, https://www.nal.usda.gov/afsic/community -supported-agriculture.

3. U.S. Department of Agriculture, National Agricultural Library, Alternative Farming Systems Information Center, *1993 Community Supported Agriculture (CSA): An Annotated Bibliography and Resource Guide* (September 1993), https://naldc.nal.usda.gov/download/6999184/PDF; Angie Vasquez et al., "Community-Supported Agriculture as a Dietary and Health Improvement Strategy: A Narrative Review," *Journal of the Academy of Nutrition and Dietetics* 117, no. 1 (2017): 83–94.

4. U.S. Department of Agriculture, 2012 Census of Agriculture, *Direct Farm Sales of Food* (2016), https://www.nass.usda.gov/Publications/Highlights/2016 /LocalFoodsMarketingPractices_Highlights.pdf.

5. Email on file with the author. For a similar firsthand account, see Robyn Van En, "Eating for Your Community: A Report from the Founder of Community Supported Agriculture," *In Context* (Fall 1995): 29, https://www.context.org /iclib/ic42/vanen/.

6. Betty T. Izumi et al., "Feasibility of Using a Community-Supported Agriculture Program to Increase Access to and Intake of Vegetables among Federally Qualified Health Center Patients," *Journal of Nutrition Education and Behavior* 50, no. 3 (2017): 289–94, https://www.jneb.org/article/S1499-4046(17)30899-0/fulltext.

7. Jack P. Cooley and Daniel A. Lass, "Consumer Benefits from Community Supported Agriculture Membership," *Review of Agricultural Economics* 20, no. 1 (1998): 227–37, https://onlinelibrary.wiley.com/doi/abs/10.2307/1349547.

8. Lydia Oberholtzer, *Community Supported Agriculture in the Mid-Atlantic Region* (Small Farm Success Project, July 2004), 23, https://www.scribd.com /document/7806331/community-supported-agriculture-in-the-mid-atlantic -region.

9. Ibid.

10. Leia M. Minaker et al., "Food Purchasing from Farmers' Markets and Community-Supported Agriculture Is Associated with Reduced Weight and Better Diets in a Population-Based Sample," *Journal of Hunger & Environmental Nutrition* 9, no. 4 (2014): 485–97.

11. Vasquez et al., "Community-Supported Agriculture as a Dietary and Health Improvement Strategy."

12. Some members of the control group were also former CSA members. J. N. Cohen, S. Gearhart, and E. Garland, "Community Supported Agriculture:

A Commitment to a Healthier Diet," *Journal of Hunger & Environmental Nutrition* 7, no. 1 (2012): 20–37, https://www.tandfonline.com/doi/abs/10.1080/19320248.2012.651393.

13. Izumi et al., "Feasibility of Using a Community-Supported Agriculture Program."

14. "The Hidden Costs of Industrial Agriculture," Union of Concerned Scientists, last modified August 24, 2008, https://www.ucsusa.org/food_and_agriculture/our-failing-food-system/industrial-agriculture/hidden-costs-of-industrial.html; David Pimentel and Michael Burgess, "Soil Erosion Threatens Food Production," *Agriculture* 3 (2013): 443–63, https://doi.org/10.3390/agriculture3030443; "Modern Agriculture: Its Effects on the Environment," *Pesticide Management Education Program: Pesticide Safety Education Program*, accessed September 2, 2019, https://ecommons.cornell.edu/handle/1813/3909.

15. Nielsen, "Unpacking the Sustainability Landscape," November 9, 2018; Louise Luttikholt and Dr. Helga Willer, "Global Organic Area Continues to Grow," International Federation of Organic Agriculture Movements, October 20, 2020, https://www.ifoam.bio/en/news/2019/02/13/world-organic-agriculture-2019.

16. Laura Reiley, "At Tampa Bay Farm-to-Table Restaurants, You're Being Fed Fiction," *Tampa Bay Times*, April 13, 2016, https://www.tampabay.com/projects/2016/food/farm-to-fable/restaurants/.

17. John C. Coffee Jr., *Gatekeepers: The Professions and Corporate Governance* (New York: Oxford University Press, 2006).

18. Eva-Marie Meemken and Matin Qaim, "Organic Agriculture, Food Security, and the Environment," *Annual Review of Resource Economics* 10 (2018): 39–63.

19. Michael Pollan, *The Omnivore's Dilemma: A Natural History of Four Meals* (New York: Penguin, 2006), 137.

20. Debroah Debord, "One Woman, One Story," *Bounty from the Box*, April 23, 2019, https://bountyfromthebox.com/one-woman-one-story/.

21. See, e.g., Izumi et al., "Feasibility of Using a Community-Supported Agriculture Program to Increase Access."

22. Jack P. Cooley, "Community Sponsored Agriculture: A Study of Shareholders' Dietary Patterns, Food Practices and Perceptions of Farm Membership," M.S. thesis, University of Massachusetts (1996), Table 5.

23. "Is CSA Right for You?" Mile Creek Farm (blog), March 29, 2018, in comments, https://milecreekfarm.com/2018/03/29/is-csa-right-for-you/.

24. Bigbirney, "America's Fragile Food Supply Chain, Part 1," Medium (blog), August 8, 2014, https://medium.com/homeland-security/americas-fragile-food-supply-chain-e387e86a355a.

25. Cooley and Lass, "Consumer Benefits from Community Supported Agriculture Membership."

26. Mary Holz-Clause, "Understanding Community Supported Agriculture," Agricultural Marketing Resource Center, 2009, https://www.agmrc.org/business-development/operating-a-business/direct-marketing/articles/understanding-community-supported-agriculture.

27. For a summary of the literature and its limitations, see Vasquez et al., "Community-Supported Agriculture as a Dietary and Health Improvement Strategy."

28. Ibid.

29. Ibid.

30. Kate Munning, "6 Things I Learned When I Joined a CSA," Community Supported Gardens at Genesis Farm, March 15, 2018, http://csgatgenesisfarm .com/6-things-i-learned-when-i-joined-a-csa/.

31. Gretchen Rubin, *The Happiness Project: Or, Why I Spent a Year Trying to Sing in the Morning, Clean My Closets, Fight Right, Read Aristotle, and Generally Have More Fun* (New York: HarperCollins, 2009).

CHAPTER 3: THE RETAIL BEHEMOTHS

1. Stephen P. Bradley, Pankaj Ghemawat, and Sharon Foley, "Wal-Mart Stores, Inc.," Harvard Business School Case 794–024, January 1994 (revised November 2002).

2. Jerry Hausman and Ephraim Leibtag, "Consumer Benefits from Increased Competition in Shopping Outlets: Measuring the Effect of Wal-Mart," *Journal of Applied Econometrics* 22, no. 7 (Dec. 2007): 1157–77. See also Emek Basker and Michael Noel, "The Evolving Food Chain: Competitive Effects of Wal Mart's Entry into the Supermarket Industry," *Journal of Economics & Management Strategy* 18, no. 4 (Winter 2009): 977–1009.

3. Hausman and Leibtag, "Consumer Benefits," 1166. For further evidence of Walmart's impact on food pricing, see N. Currie and A. Jain, *Supermarket Pricing Survey* (UBS Warburg Global Equity Research, 2002).

4. David Atkin, Benjamin Faber, and Marco Gonzalez-Navarro, "Retail Globalization and Household Welfare: Evidence from Mexico," *Journal of Political Economy* 126, no. 1 (2018): 1–73.

5. Business Planning Solutions, Global Insight Advisory Services Division, *The Price Impact of Wal-Mart: An Update Through 2006*, Global Insight Study (September 4, 2007), http://www.rossputin.com/blog/media/WalMartSept 2007.pdf.

6. Sam Walton with John Huey, *Made in America: My Story* (New York: Doubleday, 1992), 75.

7. Ibid., 51 (quoting Clarence Leis).

8. Ibid., 64.

9. Ibid., 57.

10. Ibid., 50.

11. "Fortune 500," *Fortune*, accessed October 23, 2021, https://fortune.com /fortune500/.

12. "Global 500," *Fortune*, accessed October 23, 2021, https://fortune.com/global 500/.

13. Laura Northrup, "You Probably Live Near a Walmart, So It's Depending on In-Store Pickup for Growth," *Consumerist*, August 17, 2017, https://consumeris t.com/2017/08/17/you-probably-live-near-a-walmart-so-its-depending-on-in -store-pickup-for-growth/.

14. Lisa Biank Fasig and Dan Monk, "With Wal-Mart, a Love-Hate Relationship," *Cincinnati Business Courier*, June 21, 2004.

15. Charles Fishman, *The Wal-Mart Effect: How the World's Most Powerful Company Really Works—And How It's Transforming the American Economy* (New York: Penguin Press, 2006), 162.

16. Ibid., 160–63; Fasig and Monk, "With Wal-Mart."

17. Fishman, *The Wal-Mart Effect*, 162.

18. Scott C. Friend and Patricia H. Walker, "Welcome to the New World of Merchandising," *Harvard Business Review*, November 2001, https://hbr.org/2001/11/welcome-to-the-new-world-of-merchandising.

19. Ramon Casadesus-Masanell, Eric Van Den Steen, and Karen Elterman, "The Rise and Rise (?) of Walmart (A): Battling Kmart," Harvard Business School Case 718–431, January 2018 (revised October 2018).

20. Walton and Huey, *Made in America*, xx (quoting Lou Pritchett).

21. Christopher Matthews, "10 Ways Walmart Changed the World," *Time*, June 29, 2012, https://business.time.com/2012/07/02/ten-ways-walmart-changed-the-world/slide/supplier-partnerships/.

22. Walton and Huey, *Made in America*, xx.

23. Ibid., 209.

24. Ibid., 211.

25. P. Fraser Johnson and Ken Mark, "Walmart: Supply Chain Management," Harvard Business School Case, July 2019.

26. Darrell K. Rigby, "The Future of Shopping," *Harvard Business Review*, December 2011, https://hbr.org/2011/12/the-future-of-shopping.

27. Jessica Young, "US Ecommerce Sales Grow 14.9% in 2019," *Digital Commerce 360*, February 19, 2020, https://www.digitalcommerce360.com/article/us-ecommerce-sales/.

28. Krista Garcia, "More Product Searches Start on Amazon," *eMarketer*, September 7, 2018, https://www.emarketer.com/content/more-product-searches-start-on-amazon.

29. eMarketer Editors, "Do Most Searches Really Start on Amazon?," *eMarketer*, January 7, 2020, https://www.emarketer.com/content/do-most-searchers-really-start-on-amazon.

30. "Investigation of Competition in Digital Markets," Subcommittee on Antitrust, Commercial and Administrative Law of the Committee on the Judiciary: Majority Staff Report and Recommendations (2020).

31. Sarah Perez, "Walmart Hires Former Google, Microsoft and Amazon Exec Suresh Kumar as New CTO and CDO," *Tech Crunch*, May 28, 2019, https://techcrunch.com/2019/05/28/walmart-hires-former-google-microsoft-and-amazon-exec-suresh-kumar-as-new-cto/.

32. Press release, "Amazon Unveils Its Eighth Generation Fulfillment Center," Amazon, December 1, 2014, https://press.aboutamazon.com/news-releases/news-release-details/amazon-unveils-its-eighth-generation-fulfillment-center.

33. Nick Wingfield, "As Amazon Pushes Forward with Robots, Workers Find New Roles," *New York Times*, September 10, 2017, https://www.nytimes.com/2017/09/10/technology/amazon-robots-workers.html.

34. Sean Kates, Jonathan M. Ladd, and Joshua Tucker, "Should You Worry about American Democracy? Here's What Our New Poll Finds," *Washington Post*, October 24, 2018, https://www.washingtonpost.com/news/monkey-cage/wp/2018/10/24/should-you-worry-about-american-democracy-heres-what

-our-new-poll-finds/ (summary of the major findings by the three researchers responsible for the survey).

35. Morning Consult, *Report Preview: The State of Consumer Trust 2020*, https://morningconsult.com/form/brands-well-trusted/.

36. Ganda Suthivarakom, "Welcome to the Era of Fake Products," *New York Times*, Wirecutter blog, February 11, 2020, https://www.nytimes.com/wirecutter/blog/amazon-counterfeit-fake-products/.

37. For an overview of some of this literature, see Robert Allen King, Pradeep Racherla, and Victoria D. Bush, "What We Know and Don't Know About Online Word-of-Mouth: A Review and Synthesis of the Literature," *Journal of Interactive Marketing* 28, no. 3 (August 2014): 167–83.

38. Jeff Bezos, "2018 Letter to Shareholders," Day One: The Amazon Blog, April 11, 2019, https://blog.aboutamazon.com/company-news/2018-letter-to-shareholders.

39. Brad Stone, *The Everything Store: Jeff Bezos and the Age of Amazon* (Boston: Little, Brown, 2013), 115.

40. Paavo Ritala, Arash Golnam, and Alain Wegmann, "Coopetition-based Business Models: The Case of Amazon.com," *Industrial Marketing Management* 43, no. 2 (February 2014): 236–49.

41. Doug Stephens, *Reengineering Retail: The Future of Selling in a Post-Digital World* (Figure 1 Publishing, 2017), 17.

42. Nat Levy, "New Survey Estimates Amazon Prime Membership in the U.S. Exceeds 100M," *GeekWire*, January 17, 2019, geekwire.com/2019/new-survey-estimates-amazon-prime-membership-u-s-exceeds-100m/.

43. Kaya Yurieff, "Everything Amazon Has Added to Prime Over the Years," CNN Business, April 28, 2018, https://money.cnn.com/2018/04/28/technology/amazon-prime-timeline/index.html.

44. Eugene Kim, "Amazon Can Already Ship to 72 Percent of US Population Within a Day, This Map Shows," CNBC, May 5, 2019, https://www.cnbc.com/2019/05/05/amazon-can-already-ship-to-72percent-of-us-population-in-a-day-map-shows.html.

45. Jason Newman, "Taylor Swift Brings Spectacle, Avoids Controversy at Amazon Music Concert," *Rolling Stone*, July 11, 2019, https://www.rollingstone.com/music/music-news/taylor-swift-amazon-prime-music-concert-857786/.

46. Press release, "Alexa, How Was Prime Day? Prime Day 2019 Surpassed Black Friday and Cyber Monday Combined Worldwide," Day One: The Amazon Blog, July 17, 2019, https://press.aboutamazon.com/news-releases/news-release-details/alexa-how-was-prime-day-prime-day-2019-surpassed-black-friday-0.

47. Ben Otto and Sebastian Herrera, "Amazon to Hire 100,000 in U.S. and Canada," *Wall Street Journal*, September 14, 2020, https://www.wsj.com/articles/amazon-to-hire-100–000-in-u-s-and-canada-11600071208.

48. Ibid.

49. "How Many People Work at Walmart?," Walmart Corporate—Ask Walmart, last modified March 1, 2021, https://corporate.walmart.com/askwalmart/how-many-people-work-at-walmart.

CHAPTER 4: HELPING PEOPLE BUY HOMES

1. "75th Anniversary of the Wagner-Steagall Housing Act of 1937," Franklin D. Roosevelt Presidential Library and Museum, accessed October 27, 2021, https://www.fdrlibrary.org/housing.

2. C. Lowell Harriss, *History and Policies of the Home Owners' Loan Corporation*, 1st ed. (National Bureau of Economic Research, 1951).

3. Nick Routley, "How the Composition of Wealth Differs, from the Middle Class to the Top 1%," *Visual Capitalist*, May 8, 2019, https://www.visualcapitalist .com/composition-of-wealth/; Edward N. Wolff, "Household Wealth Trends in the United States, 1962 to 2016: Has Middle Class Wealth Recovered?" NBER, NBER Working Paper No. 24085, November 2017, https://www.nber.org /papers/w24085.

4. See Brian J. McCabe, "Are Homeowners Better Citizens? Homeownership and Community Participation in the United States," *Social Forces* 91, no. 3 (2013): 929, https://doi.org/10.1093/sf/sos185.

5. Adam J. Levitin and Susan M. Wachter, *The Great American Housing Bubble: What Went Wrong and How We Can Protect Ourselves in the Future* (Cambridge, MA: Harvard University Press, 2020).

6. Ibid., 2.

7. Ibid., 24.

8. "Home Ownership Rate in the United States: 1890–2010," United States Census Bureau, October 26, 2012, census.gov/newsroom/cspan/construction _newsales/20121026_cspan_construction_newsales_slides_2.pdf.

9. Kenneth J. Robinson, "Savings and Loan Crisis: 1980–1989," Federal Reserve History, November 22, 2013, https://www.federalreservehistory.org/essays /savings_and_loan_crisis.

10. Andreas Fuster and James Vickery, "Securitization and the Fixed-Rate Mortgage," Federal Reserve Bank of New York, Staff Report No. 593, January 2013, revised June 2014, https://www.newyorkfed.org/medialibrary/media/research/ staff_reports/sr594.pdf.

11. Andreas Fuster and James Vickery, "Securitization and the Fixed-Rate Mortgage," *Review of Financial Studies* 28, no. 1 (2015): 176–211.

12. Gary Gorton and George Pennacchi, "Financial Intermediaries and Liquidity Creation," *Journal of Finance* 45, no. 1 (March 1990): 49–71.

13. Richard J. Rosen, "The Role of Securitization in Mortgage Lending," Federal Reserve Bank of Chicago, Chicago Fed Letter Number 244, November 2007, https://www.chicagofed.org/~/media/publications/chicago-fed-letter/2007 /cflnovember2007–244-pdf.pdf.

14. "Homeownership Rate for the United States," Federal Reserve Bank of St. Louis, July 28, 2020, https://fred.stlouisfed.org/series/RHORUSQ156N.

15. "Homeownership Rate for the United States: Hispanic or Latino," Federal Reserve Bank of St. Louis, July 28, 2020, https://fred.stlouisfed.org/series /HOLHORUSQ156N; "Homeownership Rate for the United States: Black or African American Alone," Federal Reserve Bank of St. Louis, July 28, 2020, https://fred.stlouisfed.org/series/BOAAAHORUSQ156N.

CHAPTER 5: THE MIDDLEMEN BEHIND THE MIDDLEMAN

1. Crown Crafts Presentation, Southwest IDEAS Investor Conference, Dallas, November 20, 2019, https://d1io3yog0oux5.cloudfront.net/_a4fb27aa28206781 e305c4067830d765/crowncrafts/db/356/3243/presentation/2019-11-20 +CrownCrafts_Presentation+-+Southwest+IDEAS.pdf.

2. Crown Crafts Annual Report on Form 10-K, for the fiscal year ended March 28, 2021, https://d1io3yog0oux5.cloudfront.net/_d15d75b05947fdf2 a5eaa86049c26345/crowncrafts/db/390/3306/annual_report/Typeset+-+Print -ready+Annual+Report.pdf.

3. "Crown Crafts, Inc. - Company Profile, Information, Business Description, History, Background Information on Crown Crafts, Inc.," Reference for Business, last visited September 9, 2020, https://www.referenceforbusiness.com /history2/68/Crown-Crafts-Inc.html.

4. Bryan Marshall, "Churchill Weavers to Close After More than 80 Years," *Richmond Register*, February 6, 2007, https://www.richmondregister.com/archives /churchill-weavers-to-close-after-more-than-years/article_e7e6675c-a8d5 -5af4-ae23-9f1884bb0c1e.html; Maggie Leininger, "Handcrafted for Success: The Churchill Weavers Collection," Kentucky Historical Society, last visited September 10, 2020, https://history.ky.gov/2017/08/21/handcrafted-success -churchill-weavers-collection/.

5. Press release, "Mohawk Industries, Inc. Completes Purchase of Assets from Crown Craft's Wovens Division," Mohawk Industries Inc., November 14, 2000, https://www.sec.gov/Archives/edgar/data/851968/000095016800002449 /0000950168-00-002449.txt; "Crown Crafts Has a Good Year," *Calhoun Times Heritage Edition*, February 24, 1988, 15.

6. In 2017, Crown Crafts acquired another company that still had U.S. manufacturing, but it then closed down that company and all of its manufacturing operations four years later. Kristen Mosbrucker, "Gonzales Children's Products Maker Shuts Down Georgia Manufacturing Hub," *Advocate*, May 11, 2021, https://www.theadvocate.com/baton_rouge/news/business/article_2be79c8e -b268-11eb-acaa-cbf9da33f514.html.

7. Marshall, "Churchill Weavers to Close After More than 80 Years."

8. Adam Smith, *The Wealth of Nations* (1776; Wordsworth Editions, 2012), 26.

9. "Ford Motor Company," *Encyclopaedia Britannica*, last modified May 18, 2020, https://www.britannica.com/topic/Ford-Motor-Company.

10. Esteban Ortiz-Ospina, "Is Globalization an Engine of Economic Development?," Our World in Data, August 1, 2017, https://ourworldindata.org /is-globalization-an-engine-of-economic-development.

11. Richard Baldwin, "Trade and Industrialisation After Globalisation's 2nd Unbundling: How Building and Joining a Supply Chain Are Different and Why It Matters," National Bureau of Economic Research, Working Paper No. 17716, https://www.nber.org/papers/w17716.pdf, 2–6.

12. Ibid., 12.

13. Banking Strategist, *Bank Merger Trends*, https://www.bankingstrategist.com /bank-merger-trends (using call report data from the FDIC).

14. See Governor Randall S. Kroszner, Member, Federal Reserve Board of Governors, *Community Banks: The Continuing Importance of Relationship Finance* (March 5, 2007), https://www.federalreserve.gov/newsevents/speech/kroszner 20070305a.htm, and sources cited therein.

15. Hubert P. Janicki and Edward Simpson Prescott, "Changes in the Size Distribution of U.S. Banks: 1960–2005," *Economic Quarterly* 92 (2006): 291–316.

16. Alicia Phaneuf, "Here Is a List of the Largest Banks in the United States by Assets in 2020," *Business Insider*, August 26, 2019, https://www.businessinsider.com/largest-banks-us-list.

17. Citigroup Inc., Annual Report (Form 10-K) 32 (February 23, 2007), https://www.sec.gov/Archives/edgar/data/831001/000119312507038505/d10k.htm.

18. "Complaint against Goldman Sachs and Fabrice Tourre," U.S. Securities and Exchange Commission, April 15, 2010, https://www.sec.gov/litigation/complaints/2010/comp21489.pdf.

19. Cezary Podkul and Megumi Fujikawa, "How a Japanese Rice Farmer Got Tangled Up in the Hertz Bankruptcy," *Wall Street Journal*, November 5, 2020, https://www.wsj.com/articles/how-a-japanese-rice-farmer-got-tangled-up -in-the-hertz-bankruptcy-11604572206?st=dnna1qut7mmziun&reflink=article _email_share.

20. Ronald J. Gilson and Jeffery N. Gordon, "The Agency Costs of Agency Capitalism: Activist Investors and the Revaluation of Governance Rights," *Columbia Law Review* 113 (2013): 863, 874.

21. Preqin, *Preqin Special Report: Private Equity Funds of Funds* 3 (November 2017), https://docs.preqin.com/reports/Preqin-Special-Report-Private-Equity -Funds-of-Funds-November-2017.pdf.

22. "Top 100 Mutual Fund Companies Ranked by AUM," *Mutual Fund Directory*, last modified August 11, 2020, https://mutualfunddirectory.org/.

23. A growing body of evidence suggests that this concentration may have indirect harms, changing the behavior of the companies in which they invest in ways that are harmful to consumers. For an overview, see Matthew Backus, Christopher Conlon, and Michael Sinkinson, "The Common Ownership Hypothesis: Theory and Evidence," Brookings, February 5, 2019, https://www.brookings.edu /research/the-common-ownership-hypothesis-theory-and-explanation/.

24. Stephanie Vatz, "Why America Stopped Making Its Own Clothes," KQED, May 24, 2013, https://www.kqed.org/lowdown/7939/madeinamerica; USDA Economic Research Service, "Food Prices and Spending," last revised August 20, 2021, https://www.ers.usda.gov/data-products/ag-and-food-statistics -charting-the-essentials/food-prices-and-spending/.

25. Derek Thompson, "How America Spends Money: 100 Years in the Life of the Family Budget," *Atlantic*, April 5, 2012, https://www.theatlantic.com/business /archive/2012/04/how-america-spends-money-100-years-in-the-life-of-the -family-budget/255475/.

26. James J. Angel, Lawrence E. Harris, and Chester S. Spatt, "Equity Trading in the 21st Century," *Quarterly Journal of Finance* 1, no. 1 (2011): 1–53.

27. Liam O'Connell, "Value of the Leading 10 Textile Exporters Worldwide in 2019, by Country," *Statista*, August 10, 2020, https://www.statista.com /statistics/236397/value-of-the-leading-global-textile-exporters-by-country/. The other jurisdictions exported, collectively, $108 billion in textiles in 2019.

28. PlasticsEurope, *Plastics—the Facts 2019: An Analysis of European Plastics Production, Demand and Waste Data*, last visited September 10, 2020, https://www .plasticseurope.org/application/files/9715/7129/9584/FINAL_web_version _Plastics_the_facts2019_14102019.pdf, 15.

29. Wayne M. Morrison, Congressional Research Service, RL33534, "China's Economic Rise: History, Trends, Challenges, and Implications for the United States" 5 (June 25, 2019), https://fas.org/sgp/crs/row/RL33534.pdf; "GDP Growth (Annual %)—China," World Bank, last visited September 10, 2020, https://data.worldbank.org/indicator/NY.GDP.MKTP.KD.ZG?end=2019& locations=CN&start=1980.

30. David H. Autor, David Dorn, and Gordon H. Hanson, "The China Syndrome: Local Labor Market Effects of Import Competition in the United States," *American Economic Review* 103, no. 6 (2013): 2121–68.

31. Ibid.

32. Charles Fishman, *The Wal-Mart Effect: How the World's Most Powerful Company Really Works—and How It's Transforming the American Economy* (New York: Penguin Press, 2006), 104.

33. David Leonhardt, "The Amazon Customers Don't See," *New York Times*, June 15, 2021, https://www.nytimes.com/2021/06/15/briefing/amazon-warehouse -investigation.html.

CHAPTER 6: WHO DO MIDDLEMEN REALLY SERVE?

1. Board of Governors of the Federal Reserve, *Report on the Economic Well-Being of U.S. Households in 2018* (May 2019), https://www.federalreserve.gov/publications /files/2018-report-economic-well-being-us-households-201905.pdf.

2. Megan Leonhardt, "41% of Americans Would Be Able to Cover a $1,000 Emergency with Savings," CNBC, January 22, 2020, https://www.cnbc .com/2020/01/21/41-percent-of-americans-would-be-able-to-cover-1000 -dollar-emergency-with-savings.html.

3. Juliana Menasce Horowitz, Ruth Igielnik, and Rakesh Kochhar, "Trends in Income and Wealth Inequality," Pew Research Center, January 9, 2020, https://www.pewsocialtrends.org/2020/01/09/trends-in-income-and-wealth -inequality/.

4. Ibid.

5. Michael J. Graetz and Ian Shapiro, *The Wolf at the Door: The Menace of Economic Insecurity and How to Fight It* (Cambridge, MA: Harvard University Press, 2020), 9.

6. Business Planning Solutions, Global Insight Advisory Services Division, *The Price Impact of Wal-Mart: An Update Through 2006*, Global Insight Study (September 4, 2007), http://www.rossputin.com/blog/media/WalMartSept2007.pdf.

7. Kim Parker et al., "What Unites and Divides Urban, Suburban and Rural Communities," Pew Research Center, May 22, 2018, https://www.pewsocialtrends .org/2020/01/09/trends-in-income-and-wealth-inequality/.

8. Siong Hook Law and Nirvikar Singh, "Does Too Much Finance Harm Economic Growth," *Journal of Banking & Finance* 41 (April 2014): 36–44, https:// doi.org/10.1016/j.jbankfin.2013.12.020; Jean-Louis Arcand, Enrico Berkes, and

Ugo Panizza, "Too Much Finance?," *Journal of Economic Growth* 20 (2015): 105–48.

9. Robin Greenwood and David Scharfstein, "The Growth of Finance," *Journal of Economic Perspectives* 27, no. 2 (2013): 3–28, www.jstor.org/stable /23391688.

10. Rebecca Stropoli, "How the 1 Percent's Savings Buried the Middle Class in Debt," *Chicago Booth Review*, May 25, 2021, https://review.chicagobooth.edu /economics/2021/article/how-1-percent-s-savings-buried-middle-class-debt and sources cited therein.

11. Thomas Philippon, "Has the US Finance Industry Become Less Efficient? On the Theory and Measurement of Financial Intermediation," *American Economic Review* 105, no. 4 (2015): 1408–38, www.jstor.org/stable/43495423.

12. Walton and Huey, *Made in America*, 57.

13. Michelle Yan, "9 Sneaky Ways Walmart Makes You Spend More Money," *Business Insider*, January 23, 2019, https://www.businessinsider.com/how-walmart -gets-you-spend-more-money-2019-1; Áine Cain, "10 Sneaky Ways Walmart Gets You to Spend More Money," *Business Insider*, March 15, 2019, https:// www.businessinsider.com/walmart-spend-more-money-strategy-2019-3.

14. Yan, "9 Sneaky Ways Walmart Makes You Spend More Money."

15. Judith A. Chevalier, Anil K. Kashyap, and Peter E. Rossi, "Why Don't Prices Rise During Periods of Peak Demand? Evidence from Scanner Data," *American Economic Review* 93, no. 1 (2003), 15–37, http://ezproxy.cul.columbia .edu/login?url=https://www-proquest-com.ezproxy.cul.columbia.edu/docview /38447647?accountid=10226.

16. Jeremy Sporn and Stephanie Tuttle, "5 Surprising Findings About How People Actually Buy Clothes and Shoes," *Harvard Business Review*, June 6, 2018, https://hbr.org/2018/06/5-surprising-findings-about-how-people-actually-buy -clothes-and-shoes.

17. Harry Brignull, "Dark Patterns: Dirty Tricks Designers Use to Make People Do Stuff," 90 Percent of Everything (blog), July 8, 2010, https://90percentof everything.com/2010/07/08/dark-patterns-dirty-tricks-designers-use-to -make-people-do-stuff/.

18. Arushi Jaiswal, "Dark Patterns in UX: How Designers Should Be Responsible for Their Actions," Medium, April 15, 2018, https://uxdesign.cc/dark-patterns -in-ux-design-7009a83b233c.

19. Ibid., 81:2.

20. Ibid.

21. Roman Chuprina, "Artificial Intelligence for Retail in 2020: 12 Real-World Use Cases," SPD Group, December 20, 2019, https://spd.group/artificial -intelligence/ai-for-retail/.

22. Matt Smith, "Walmart's New Intelligent Retail Lab Shows a Glimpse into the Future of Retail, IRL," Walmart, April 25,2019, https://corporate.walmart. com/newsroom/2019/04/25/walmarts-new-intelligent-retail-lab-shows-a -glimpse-into-the-future-of-retail-irl.

23. Joseph Turow, *The Aisles Have Eyes: How Retailers Track Your Shopping, Strip Your Privacy, and Define Your Power* (New Haven, CT: Yale University Press, 2017), 3.

24. Erik Brynjolfsson and Andrew McAfee, "How AI Fits into Your Science Team," *Harvard Business Review*, July 21, 2017, https://starlab-alliance.com /wp-content/uploads/2017/09/The-Business-of-Artificial-Intelligence.pdf; Cathy O'Neil, *Weapons of Math Destruction: How Big Data Increases Inequality and Threatens Democracy* (New York: Crown, 2016).

25. Brent R. Smith and Greg Linden, "Two Decades of Recommender Systems at Amazon.com," *IEEE Internet Computing* 21, no. 3 (2017): 12–18, https://doi .org/10.1109/MIC.2017.72.

26. Zahy Ramadan et al., "Fooled in the Relationship: How Amazon Prime Members' Sense of Self-Control Counter-intuitively Reinforces Impulsive Buying Behavior," *Journal of Consumer Behaviour* (May 2021).

27. Ibid.

28. Nat Levy, "New Survey Estimates Amazon Prime Membership in the U.S. Exceeds 100M," *GeekWire*, January 17, 2019, geekwire.com/2019/new-survey -estimates-amazon-prime-membership-u-s-exceeds-100m/.

29. C. Courtemanche and A. Carden, "Supersizing Supercenters? The Impact of Walmart Supercenters on Body Mass Index and Obesity," *Journal of Urban Economics* 69, no. 2 (March 2011): 165–81, https://doi.org/10.1016/j .jue.2010.09.005.

30. Ibid., 166.

31. Ibid.

32. Floriana S. Luppino et al., "Overweight, Obesity, and Depression: A Systematic Review and Meta-analysis of Longitudinal Studies," *Archives of General Psychiatry* 67, no. 3 (2010): 220–29.

33. Stephanie Vatz, "Why America Stopped Making Its Own Clothes," KQED, May 24, 2013, https://www.kqed.org/lowdown/7939/madeinamerica.

34. "UN Alliance Aims to Put Fashion on Path to Sustainability," UNECE, July 12, 2018, https://www.unece.org/info/media/presscurrent-press-h/forestry -and-timber/2018/un-alliance-aims-to-put-fashion-on-path-to-sustainability /doc.html.

35. Nicholas Gilmore, "Ready-to-Waste: America's Clothing Crisis," *Saturday Evening Post,* January 16, 2018, https://www.saturdayeveningpost.com /2018/01/ready-waste-americas-clothing-crisis/.

36. "Textiles: Material-Specific Data," U.S. Environmental Protection Agency, modified October 7, 2020, https://www.epa.gov/facts-and-figures-about -materials-waste-and-recycling/textiles-material-specific-data.

37. "Books: Best Sellers: Advice, How-To & Miscellaneous," *New York Times*, May 12, 2019, https://www.nytimes.com/books/best-sellers/2019/05/12 /advice-how-to-and-miscellaneous/.

38. Jamie Feldman, "I Got Rid of Half My Wardrobe Using Marie Kondo's Methods. Here's What I Learned," *HuffPost*, January 9, 2019, https://www.huffpost.com /entry/marie-kondo-clothes-tidying-method_n_5c2e5fc1e4b05c88b70755ad; Gabrielle Savoie, "A Marie Kondo Expert Donated 90% of My Wardrobe," MyDomaine, July 14, 2019, https://www.mydomaine.com/marie-kondo-method.

39. Gretchen Rubin, *The Happiness Project: Or, Why I Spent a Year Trying to Sing in the Morning, Clean My Closets, Fight Right, Read Aristotle, and Generally Have More Fun* (New York: HarperCollins, 2009).

40. Jeffrey Dew, Sonya Britt, and Sandra Huston, "Examining the Relationship Between Financial Issues and Divorce," *Family Relations: Interdisciplinary Journal of Applied Family* 61, no. 4 (October 2012): 615–28.

41. Brienne Walsh, "This Is 'One of the Largest and Most Painful Issues' Couples Deal With," Millie, January 21, 2021, https://www.synchronybank.com/blog/millie/the-price-of-love/.

42. Charles Fishman, *The Wal-Mart Effect: How the World's Most Powerful Company Really Works—and How It's Transforming the American Economy* (New York: Penguin Press, 2006), 258.

43. Ibid., 220.

44. Jonathan Hancock, "My Love-Hate Relationship with Amazon," MindTools, March 11, 2021, https://www.mindtools.com/blog/my-love-hate-relationship-with-amazon/.

45. Brian Dumaine, "Even Americans Who Hate Amazon Can't Seem to Live Without It," Literary Hub, May 19, 2020, https://lithub.com/even-americans-who-hate-amazon-cant-seem-to-live-without-it/ (citing a poll conducted by the Max Borges Agency).

46. Chang-Tai Hsieh, and Enrico Moretti, "Can Free Entry Be Inefficient? Fixed Commissions and Social Waste in the Real Estate Industry," *Journal of Political Economy* 111, no. 5 (October 2003): 1076–1121.

47. Ibid., 1089.

48. Kriston McIntosh et al., "Examining the Black-White Wealth Gap," Brookings, February 27, 2020, https://www.brookings.edu/blog/up-front/2020/02/27/examining-the-black-white-wealth-gap/.

49. Mehrsa Baradaran, *The Color of Money: Black Banks and the Racial Wealth Gap* (Cambridge, MA: Harvard University Press, 2017); William R. Emmons, "Housing Wealth Climbs for Hispanics and Blacks, Yet Racial Wealth Gaps Persist," Federal Reserve Bank of St. Louis, April 1, 2020, https://www.stlouisfed.org/publications/housing-market-perspectives/2020/racial-wealth-gaps-persist.

50. Joint Center for Housing Studies of Harvard University, *The State of the Nation's Housing 2018* (2018), https://www.jchs.harvard.edu/sites/default/files/reports/files/Harvard_JCHS_State_of_the_Nations_Housing_2018.pdf, 3.

51. Alanna McCargo and Sarah Strochak, "Mapping the Black Ownership Gap," Urban Wire: Housing and Housing Finance, February 26, 2018, https://www.urban.org/urban-wire/mapping-black-homeownership-gap.

52. Andre M. Perry, Jonathan Rothwell, and David Harshbarger, "The Devaluation of Assets in Black Neighborhoods," Brookings, November 27, 2018, https://www.brookings.edu/research/devaluation-of-assets-in-black-neighborhoods/.

53. Keeanga-Yamahtta Taylor, *Race for Profit: How Banks and the Real Estate Industry Undermined Black Homeownership* (Chapel Hill: University of North Carolina Press, 2019).

54. A number of other researchers have also examined these dynamics. See Rose Helper, *Racial Policies and Practices of Real Estate Brokers* (Minneapolis: University of Minnesota Press, 1969), 143–54; Dmitri Mehlhorn, "A Requiem for Blockbusting: Law, Economics, and Race-Based Real Estate Speculation," *Fordham Law Review* 67, no. 3 (1998): 1145, 1176–79, https://ir.lawnet.fordham.edu/cgi/viewcontent.cgi?article=3528&context=flr.

55. Jacob W. Faber, "Racial Dynamics of Subprime Mortgage Lending at the Peak," *Housing Policy Debate* 23, no. 2 (2013): 328–49, https://doi.org/10.1080/10511 482.2013.771788.

56. Ibid., 343.

CHAPTER 7: MIDDLEMEN PERPETUATING THE NEED FOR MIDDLEMEN

1. Jeff Mindham, "8 Real Estate Startups That Crashed and Burned (Or Simply Sputtered Out)," Disruptor, November 22, 2017, https://www.disruptordaily. com/8-real-estate-startups-crashed-burned-simply-sputtered/.

2. Michele Lerner, "Commissions of 6 Percent for Home Sales Once Were the Norm. That's Changing.," *Washington Post*, April 15, 2016, https://www .washingtonpost.com/realestate/commissions-of-6-percent-for-home-sales -are-the-norm-but-that-is-changing/2016/04/13/91bb758c-fb55-11e5-886f-a03 7dba38301_story.html.

3. Transcript, *What's New in Residential Real Estate Brokerage Competition— An FTC-DOJ Workshop* (June 5, 2018), https://www.ftc.gov/system/files /documents/videos/whats-new-residential-real-estate-brokerage-competition -part-1/ftc-doj_residential_re_brokerage_competition_workshop_transcript _segment_1.pdf.

4. *Realcomp II, Ltd. v. FTC*, 635 F. 3d 815 (6th Cir. 2011).

5. *Freeman v. San Diego Ass'n of Realtors*, 322 F.3d 1133 (9th Cir. 2003).

6. Jean-Charles Rochet, "Two-Sided Markets: A Progress Report," *RAND Journal of Economics* 37, no. 3 (2006): 645–67.

7. Staff of House Committee on the Judiciary, 116th Congress, *Investigation of Competition of Digital Markets: Majority Staff Report and Recommendations* (Comm. Print 2020), https://int.nyt.com/data/documenttools/house-antitrust -report-on-big-tech/b2ec22cf340e1af1/full.pdf.

8. Brian Maass, "'Your Business Model Sucks: I Hope It Burns': Fear & Loathing in Real Estate," CBS Denver, May 22, 2017, https://denver.cbslocal .com/2017/05/22/trelora-real-estate-controversy/.

9. Transcript, *What's New in Residential Real Estate Brokerage Competition*.

10. Steven D. Levitt and Chad Syverson, "Antitrust Implications of Home Seller Outcomes When Using Flat-Fee Real Estate Agents," *Brookings-Wharton Papers on Urban Affairs* (2008): 47–93.

11. PwC, "Considering an IPO? First, Understand the Costs," as last modified March 17, 2021, https://www.pwc.com/us/en/services/deals/library/cost-of -an-ipo.html.

12. Minmo Gahng, Jay R. Ritter, and Donghang Zhang, "SPACs" (working paper, revised July 2021), https://ssrn.com/abstract=3775847.

13. Ibid.; Michael Klausner, Michael Ohlrogge, and Emily Ruan, "A Sober Look at SPACs," *Yale Journal on Regulation*, forthcoming.

14. Tim Jenkinson and Howard Jones, "IPO Pricing and Allocation: A Survey of the Views of Institutional Investors," *Review of Financial Studies* 22, no. 4 (April 2009): 1477–1504, https://doi.org/10.1093/rfs/hhn079; Michael A. Goldstein, Paul Irvine, and Andy Puckett, "Purchasing IPOs with Commissions,"

Journal of Financial and Quantitative Analysis 46, no. 5 (October 2011): 1193–225.

15. Michelle Lowry, Roni Michaely, and Ekaterina Volkova, "Initial Public Offerings: A Synthesis of the Literature and Directions for Future Research," *Foundations and Trends in Finance* 11, nos. 3–4 (2017): 154–320, http://dx.doi .org/10.1561/0500000050.

16. See Chapter 5.

17. Committee on the Judiciary, *Investigation of Competition of Digital Markets* and sources cited therein.

18. Laura Stevens and Sara Germano, "Nike Thought It Didn't Need Amazon— Then the Ground Shifted," *Wall Street Journal*, June 28, 2017, https://www.wsj .com/articles/how-nike-resisted-amazons-dominance-for-years-and-finally -capitulated-1498662435.

19. C. Scott Hemphill and Tim Wu, "Nascent Competitors," *University of Pennsylvania Law Review* 168 (2021): 1879–1910.

20. Brad Stone, *Amazon Unbound: Jeff Bezos and the Invention of a Global Empire* (New York: Simon & Schuster, 2021), 222–24.

21. Amy Klobuchar, *Taking on Monopoly Power from the Gilded Age to the Digital Age* (New York: Knopf, 2021); Lina M. Khan, "Amazon's Antitrust Paradox," *Yale Law Journal* 126, no. 3 (2017): 710–805; Tim Wu, *The Curse of Bigness: Antitrust in the New Gilded Age* (Columbia Global Reports, 2018); Teachout, *Break 'Em Up*.

22. Ala. Code Sec. 34–27–36; Alaska Stat. Sec. 08.88.401; Iowa Code Sec. 543B.60A; Kan. Stat. Ann. Sec. 58–3062 (repealed); KRS 324.160; La. Rev. Stat. Ann. Sec. 37:1455; Miss. Code Ann. Sec. 73–35–21; Mo. Rev. Stat. Sec. 339.150; Montana Board of Realty Regulation R. 24.210.641(5) (repealed); N.H. Rev. Stat. Ann. Sec. 331-A:26(XXIV) (repealed); N.J. Stat. Ann. Sec. 45:15–3.1 (repealed); Century Code 43–23–11–1.1 (repealed); Okla. Stat. Ann. tit. 59, Sec. 858–312; Or. Rev. Stat. Sec. 696.290(1); South Carolina Code Sec. 40–57–145(11) (repealed); South Dakota Real Estate Commission Resolution 06–30–05–01 (repealed); Tenn. Code. Ann. Sec. 62–13–302; Legislative Rule CSR Sec. 174–1–11.11.1 (repealed).

23. John M. de Figueiredo and Brian Kelleher Richter, "Advancing the Empirical Research on Lobbying," *Annual Review of Political Science* 17, no. 1 (2014): 163–85.

24. Ibid.

25. Inman, "A Response to a Broker's Open Letter from NAR CEO Dale Stinton," Inman, February 2, 2017, https://www.inman.com/2017/02/02/a-response-to-a-brokers-open-letter-from-nar-ceo-dale-stinton/.

26. "National Assn of Realtors," OpenSecrets.org: Influence & Lobbying, Center for Responsive Politics, modified Sept. 21, 2020, https://www.opensecrets.org/ orgs/summary.php?id=D000000062&cycle=2012.

27. Inman, "A Response to a Broker's Open Letter."

28. William G. Gale, "It's Time to Gut the Mortgage Interest Deduction," *Brookings: Up Front*, November 6, 2017, https://www.brookings.edu/blog/up -front/2017/11/06/its-time-to-gut-the-mortgage-interest-deduction/.

29. "Frequently Asked Questions: Banking Conglomerates Permanently Barred from Real Estate Activities by the FY 2009 Omnibus Appropriations Act," National Association of Realtors, last visited March 20, 2021, https://www

.nar.realtor/banks_and_commerce.nsf/Pages/banks_permanently_barred
_faqs?OpenDocument.

30. Lawrence J. White, "Housing Policy, The Morning After," *Milken Institute Review: Articles*, May 2, 2016, https://www.milkenreview.org/articles/housing -policy-the-morning-after.

31. "Two Big Problems Facing Real Estate; Construction Costs and the Tax Burden Must Be Reduced, Says Secretary Nelson. Checks Home Ownership Mass Production Suggsated [*sic*] as a Cure for Constantly Increasing Building Costs," *New York Times*, February 9, 1930, https://timesmachine.nytimes.com/times machine/1930/02/09/92074592.pdf?pdf_redirect=true&ip=0.

32. "Finance/Insurance/Real Estate: Summary," OpenSecrets.org: Influence & Lobbying, Interest Groups, Center for Responsive Politics, modified September 21, 2020, https://www.opensecrets.org/industries/indus.php?ind=F.

33. "Amazon.com," OpenSecrets.org: Influence & Lobbying, Center for Responsive Politics, accessed October 3, 2020, https://www.opensecrets.org/orgs /amazon-com/lobbying?id=D000023883.

34. Naomi Nix, "Amazon Is Flooding D.C. with Money and Muscle: The Influence Game," *Bloomberg Businessweek*, March 7, 2019, https://www.bloomberg .com/graphics/2019-amazon-lobbying/?sref=0SF97H1m.

35. "Lobbying: Top Spenders," OpenSecrets.org: Influence & Lobbying, Center for Responsive Politics, modified July 23, 2020, https://www.opensecrets.org /federal-lobbying/top-spenders?cycle=2019.

36. Hans R. Stoll, *Revolution in the Regulation of Securities Markets: An Examination of the Effects of Increased Competition in Case Studies in Regulation: Revolution and Reform* (Leonard W. Weiss & Michael W. Klass, eds., 1981); William F. Baxter, "NYSE Fixed Commission Rates: A Private Cartel Goes Public," *Stanford Law Review* 22, no. 4 (1970): 675–712.

37. Gregg A. Jarrell, "Change at the Exchange: The Causes and Effects of Deregulation," *Journal of Law & Economics* 27, no. 2 (1984): 273–312; "NYSE Was Revolutionized by SEC Abolition of Fixed Commissions," *Washington Post*, July 21, 1985, https://www.washingtonpost.com/archive/business/1985/07/21 /nyse-was-revolutionized-by-sec-abolition-of-fixed-commissions/8726b8b1 –8013–4bcf-aad8–776fcc65f417/.

38. See Chapter 4.

39. At the time, there was also a second federal bank regulator, the Office of Thrift Supervision, which engaged in similar and in many ways more troubling efforts to shield institutions it oversaw from state laws.

40. Patricia A. McCoy and Kathleen C. Engel, "Federal Preemption and Consumer Financial Protection; Past and Future," *Banking & Financial Services Policy Report* 31, no. 3 (March 2012): 25–36; Michael S. Barr, Howell E. Jackson, and Margaret E. Tayhar, *Financial Regulation: Law and Policy* 2nd ed. (St. Paul, MN: Foundation Press, 2018).

41. Bank Activities and Operations; Real Estate Lending and Appraisals, 69 Fed. Reg. 1904, 1906 (January 13, 2004), https://www.govinfo.gov/content/pkg/FR -2004-01-13/pdf/04-586.pdf#page=1.

42. McCoy and Engel, "Federal Preemption."

43. Joint Center for Housing Studies of Harvard University, *The State of the Nation's Housing 2018* (2018), https://www.jchs.harvard.edu/sites/default

/files/reports/files/Harvard_JCHS_State_of_the_Nations_Housing_2018.pdf; Danielle Douglas-Gabriel, "Home Buying While Black," *Washington Post*, September 7, 2017, https://www.washingtonpost.com/realestate/home-buying -while-black/2017/09/07/133e286a-8995-11e7-a50f-e0d4e6ec070a_story.html.

44. Arthur E. Wilmarth, *Taming the Megabanks: Why We Need a New Glass-Steagall Act* (New York: Oxford University Press, 2020).

45. Commodity Futures Modernization Act, Pub. L. 106–554, 114 Stat. 2763 (2000) (codified as amended at 7 U.S.C. §§ 27–27f (2018)).

46. Wilmarth, *Taming the Megabanks*.

47. Edward R. Morrison, Mark J. Roe, and Christopher S. Sontchi, "Rolling Back the Repo Safe Harbors," *Business Lawyer* 69, no. 4 (August 2014): 1015–47.

CHAPTER 8: THE MYTH OF SUPPLY CHAIN ACCOUNTABILITY

1. Board of Governors of the Federal Reserve System, Meeting of the Federal Open Market Committee on September 18, 2007, https://www.federalreserve. gov/monetarypolicy/files/fomc20070918meeting.pdf.

2. Ibid.

3. Sudip Kar-Gupta and Yann Le Guernigou, "BNP Freezes $2.2 BLN of Funds over Subprime," Reuters, August 9, 2007, https://www.reuters.com/article /us-bnpparibas-subprime-funds/bnp-freezes-2–2-bln-of-funds-over-subprime -idUSWEB612920070809.

4. Board of Governors of the Federal Reserve System, Meeting of the Federal Open Market Committee on September 18, 2007, 3–8.

5. Daniel Covitz, Nellie Liang, and Gustavo Suarez, "The Anatomy of a Financial Crisis: The Evolution of Panic-driven Runs in the Asset-Backed Commercial Paper Market," Federal Reserve Bank of San Francisco, December 22, 2008.

6. Ibid.

7. Kathryn Judge, "Information Gaps and Shadow Banking," *Virginia Law Review* 103, no. 3 (2017): 411–80.

8. Laura Kusisto, "Many Who Lost Homes to Foreclosure in Last Decade Won't Return—NAR," *Wall Street Journal*, April 20, 2015, https://www.wsj.com /articles/many-who-lost-homes-to-foreclosure-in-last-decade-wont-return -nar-1429548640.

9. U.S. Government Accountability Office, *Financial Regulatory Reform: Financial Crisis Losses and Potential Impacts of Dodd-Frank Act* (January 2013).

10. Fabian T. Pfeffer, Sheldon Danziger, and Robert F. Schoeni, *Wealth Levels, Wealth Inequity, and the Great Recession*, Russell Sage Foundation (June 2014).

11. Gillian B. White, "The Recession's Racial Slant," *Atlantic*, June 24, 2015, https:// www.theatlantic.com/business/archive/2015/06/black-recession-housing -race/396725/.

12. Board of Governors of the Federal Reserve System, Meeting of the Federal Open Market Committee on September 18, 2007, 90.

13. Samuel G. Hanson and Adi Sunderam, "Are There Too Many Safe Securities? Securitization and the Incentives for Information Production," *Journal of Financial Economics* 108, no. 3 (2013): 565–84.

14. Senate Committee on Banking, Housing, and Urban Affairs, "Turmoil in U.S. Credit Markets: Examining the Recent Actions of Federal Financial Regulators," April 3, 2008, Senate Hearing 110–974, https://www.govinfo.gov/content/pkg/CHRG-110shrg50394/pdf/CHRG-110shrg50394.pdf, 23.

15. Andrea Riquier, "Has the Housing Market Recovered? Ask the 1.4 Million Homeowners Still Underwater," *MarketWatch*, December 14, 2017, https://www.marketwatch.com/story/has-the-housing-market-recovered-ask-the-14-million-underwater-homeowners-2017-12-13.

16. Kathryn Judge, "Fragmentation Nodes: A Study in Financial Innovation, Complexity, and Systemic Risk," *Stanford Law Review* 64, no. 3 (2012): 657–725; Anna Gelpern and Adam J. Levitin, "Rewriting Frankenstein Contracts: Workout Prohibitions in Residential Mortgage-Backed Securities," *Southern California Law Review* 82, no. 6 (2009): 1075–1152.

17. Judge, "Fragmentation Nodes."

18. Jen Wieczner, "The Case of the Missing Toilet Paper: How the Coronavirus Exposed U.S. Supply Chain Flaws," *Fortune*, May 18, 2020, https://fortune.com/2020/05/18/toilet-paper-sales-surge-shortage-coronavirus-pandemic-supply-chain-cpg-panic-buying/.

19. Mark Sweney, "Global Shortage in Computer Chips 'Reaches Crisis Point,'" *Guardian*, March 21, 2021, https://www.theguardian.com/business/2021/mar/21/global-shortage-in-computer-chips-reaches-crisis-point; Yasmin Tadjdeh, "Semiconductor Shortage Shines Light on Weak Supply Chain," *National Defense*, May 21, 2021, https://www.nationaldefensemagazine.org/articles/2021/5/21/semiconductor-shortage-shines-light-on-weak-supply-chain.

20. Alexandre Tanzi, "Used-Car Prices Are Poised to Peak in U.S. After Pandemic Surge," Bloomberg, June 24, 2021, https://www.bloomberg.com/news/articles/2021-06-24/used-car-prices-are-poised-to-peak-in-u-s-after-pandemic-surge?sref=0SF97H1m; Tom Krisher, "Some Used Vehicles Now Cost More Than Original Sticker Price," AP News, June 22, 2021, https://apnews.com/article/science-technology-prices-health-coronavirus-pandemic-bb0ebc0112b9eab606936499b5e2c6f5.

21. American Wood Council, "Wood Products Manufacturers Expand Capacity, Continue High Levels of Production," May 25, 2021, https://awc.org/news/2021/05/25/wood-products-manufacturers-expand-capacity-continue-high-levels-of-production.

22. Marc Levinson, *Outside the Box: How Globalization Changed from Moving Stuff to Spreading Ideas* (Princeton, NJ: Princeton University Press, 2020), 156.

23. Henry Ren, "Higher Shipping Costs Are Here to Stay, Sparking Price Increases," Bloomberg, April 12, 2021, https://www.bloomberg.com/news/articles/2021-04-12/higher-shipping-costs-are-here-to-stay-sparking-price-increases?sref=0SF97H1m.

24. Costas Paris, "Shipments Delayed: Ocean Carrier Shipping Times Surge in Supply-Chain Crunch," *Wall Street Journal*, May 18, 2021, https://www.wsj.com/articles/shipments-delayed-ocean-carrier-shipping-times-surge-in-supply-chain-crunch-11621373426.

25. Tracy Alloway and Joe Weisenthal, interview with Ryan Petersen, "How the World's Companies Wound Up in a Deepening Supply Chain Nightmare," *Odd Lots Podcast*, May 17, 2021, https://www.bloomberg.com/news

/articles/2021-05-17/how-the-world-s-companies-wound-up-in-a-deepening
-supply-chain-nightmare?srnd=oddlots-podcast&sref=0SF97H1m.

26. Jeanna Smialek, "Prices Jumped 5% in May from Year Earlier, Stoking Debate in Washington," *New York Times*, June 10, 2021, https://www.nytimes
.com/2021/06/10/business/consumer-price-index-may-2021.html.

27. Jeanna Smialek, "Larry Summers Warned About Inflation. Fed Officials Push Back," *New York Times*, March 25, 2021, https://www.nytimes.com
/2021/03/25/business/economy/larry-summers-federal-reserve.html.

28. *Mortgage Fraud Report 2008*, Federal Bureau of Investigation: Reports and Publications, https://www.fbi.gov/stats-services/publications/mortgage-fraud-2008.

29. Jesse Eisinger and Jake Bernstein, "The Magnetar Trade: How One Hedge Fund Helped Keep the Bubble Going," ProPublica, April 9, 2010, https://www
.propublica.org/article/all-the-magnetar-trade-how-one-hedge-fund-helped
-keep-the-housing-bubble.

30. U.S. Department of Justice, "Bank of America to Pay $16.65 Billion in Historic Justice Department Settlement for Financial Fraud Leading Up to and During the Financial Crisis," Justice News, press release 14–884 (August 21, 2014), https://www.justice.gov/opa/pr/bank-america-pay-1665-billion-historic
-justice-department-settlement-financial-fraud-leading; U.S. Department of Justice, Bank of America Corporation: Statement of Facts, Justice News, press release 14–884, Annex 1 (August 21, 2014), https://www.justice.gov/iso/opa
/resources/4312014829141220799708.pdf.

31. John A. Ruddy, Murli Rajan, and Iordanis Petsas, "A Study of RMBS Litigation Cases of Six Major U.S. Banks," *Journal of Structured Finance* 23, no. 3 (Fall 2017): 91–99, https://doi.org/10.3905/jsf.2017.23.3.091.

32. Jennifer Taub, *Big Dirty Money: The Shocking Injustice and Unseen Cost of White Collar Crime* (New York: Viking, 2020); Jesse Eisinger, *The Chickenshit Club: Why the Justice Department Fails to Prosecute* (New York: Simon & Schuster, 2018).

33. Daniel Kahneman, "Maps of Bounded Rationality: Psychology for Behavioral Economics," *American Economic Review* 93 (2003): 1449–75.

34. Katherine L. Milkman, Todd Rogers, and Max H. Bazerman, "Harnessing Our Inner Angels and Demons: What We Have Learned About Want/Should Conflicts and How That Knowledge Can Help Us Reduce Short-Sighted Decision Making," *Perspectives on Psychological Science* 3 (2008): 324–38; Hunt Allcott and Nathan Wozny, "Gasoline Prices, Fuel Economy, and the Energy Paradox," *Review of Economics and Statistics* 96 (2012): 779–95; Shahzeen Z. Attari et al., "Public Perceptions of Energy Consumption and Savings," *Proceedings of the National Academy of Sciences* 107 (2010): 16054–59; Shahzeen Z. Attari, "Perceptions of Water Use," *Proceedings of the National Academy of Sciences* 111 (2014): 5129–34.

35. Bureau of International Labor Affairs, Department of Labor, *2020 List of Goods Produced by Child Labor or Forced Labor* (September 2020), https://www.dol
.gov/sites/dolgov/files/ILAB/child_labor_reports/tda2019/2020_TVPRA
_List_Online_Final.pdf.

36. Hinrich Voss et al., "International Supply Chains: Compliance and Engagement with the Modern Slavery Act," *Journal of the British Academy* 7, no. 1 (2019): 61–76, https://doi.org/10.5871/jba/007s1.061.

37. Margaret Besheer, "At UN: 39 Countries Condemn China's Abuses of Uighurs," Voice of America, October 6, 2020.

38. "Global Supply Chains, Forced Labor, and the Xinjiang Uyghur Autonomous Region," Congressional-Executive Commission in China: Staff Research Report (March 2020).

39. "Nike, Inc. Statement on Forced Labor, Human Trafficking and Modern Slavery for Fiscal Year 2020," Nike, modified March 8, 2021, https://www.nike.com/help/a/supply-chain.

40. Reuters: Dhaka, "Rana Plaza Collapse: 38 Charged with Murder Over Garment Factory Disaster," *Guardian*, July 18, 2016, https://www.theguardian.com/world/2016/jul/18/rana-plaza-collapse-murder-charges-garment-factory.

41. Dana Thomas, "Why Won't We Learn from the Survivors of the Rana Plaza Disaster?" *New York Times*, April 24, 2018, https://www.nytimes.com/2018/04/24/style/survivors-of-rana-plaza-disaster.html.

42. "UN Alliance Aims to Put Fashion on Path to Sustainability," UNECE, July 12, 2018, https://www.unece.org/info/media/presscurrent-press-h/forestry-and-timber/2018/un-alliance-aims-to-put-fashion-on-path-to-sustainability/doc.html.

43. Cyril Villemain, "UN Launches Drive to Highlight Environmental Cost of Staying Fashionable," *UN News*, March 25, 2019, https://news.un.org/en/story/2019/03/1035161.

44. Deborah Drew and Genevieve Yehounme, "The Apparel Industry's Environmental Impact in 6 Graphics," World Resources Institute, July 5, 2017, https://www.wri.org/blog/2017/07/apparel-industrys-environmental-impact-6-graphics.

45. Villemain, "UN Launches Drive."

46. Omri Ben-Shahar and Carl E. Schneider, *More Than You Wanted to Know: The Failure of Mandated Disclosure* (Princeton, NJ: Princeton University Press, 2014).

47. 154 Cong. Rec. S1047 (daily ed., February 14, 2008) (statement of Senator Brownback), https://www.congress.gov/crec/2008/02/14/CREC-2008-02-14-pt1-PgS1047-2.pdf.

48. Senator Russ Feingold opined, "We need to finally get serious about addressing the underlying issues that make this war profitable and allow it to persist." 155 Cong. Rec. S13030 (daily ed., December 11, 2009), https://www.congress.gov/111/crec/2009/12/11/CREC-2009–12–11-pt1-PgS13030.pdf; Dodd-Frank Wall Street Reform and Consumer Protection Act, Pub. L. No. 111–203, § 1502, 124 Stat. 2213 (2010).

49. Debapratim Purkayastha and Syed Abdul Samad, "Apple and Conflict Minerals: Ethical Sourcing for Sustainability," *IUP Journal of Operations Management* 14, no. 2 (May 2015): 59–77, https://ssrn.com/abstract=2686957.

50. U.S. Government Accountability Office, GAO-15–561, "SEC Conflict Minerals Rule: Initial Disclosures Indicate Most Companies Were Unable to Determine the Source of Their Conflict Minerals," (August 2015) 15, 19, https://www.gao.gov/assets/680/672051.pdf.

51. U.S. Government Accountability Office, GAO-20-595, "Conflict Minerals: Action Needed to Assess Progress Addressing Armed Groups' Exploitation of Minerals," (September 2020) 17, 18, https://www.gao.gov/assets/710/709359.pdf.

52. Alex Brackett, Estelle Levin, and Yves Melin, "Revisiting the Conflict Minerals Rule," *Global Trade & Customs Journal* 10, no. 2 (2015): 73, 77; Lauren Wolfe, "How Dodd-Frank Is Failing Congo," *Foreign Policy*, February 2, 2015, https://foreignpolicy.com/2015/02/02/how-dodd-frank-is-failing-congo-mining-conflict-minerals/; Sudarsan Raghavan, "How a Well-Intentioned U.S. Law Left Congolese Miners Jobless," *Washington Post*, November 30, 2014, https://www.washingtonpost.com/world/africa/how-a-well-intentioned-us-law-left-congolese-miners-jobless/2014/11/30/14b5924e-69d3-11e4-9fb4-a62 2dae742a2_story.html?utm_term=.513f6c2c9c0b.

53. Laura E. Seay, "What's Wrong with Dodd-Frank 1502? Conflict Minerals, Civilian Livelihoods, and the Unintended Consequences of Western Advocacy," 15 (Center for Global Development, Working Paper No. 284, 2012), http://www.cgdev.org/content/publications/detail/1425843/.

54. Apple Inc., *Conflict Minerals Disclosure and Report 2020* (February 10, 2021), https://www.apple.com/supplier-responsibility/pdf/Apple-Conflict-Minerals-Report.pdf.

55. Andreas C. Drichoutis et al., "Consumer Preferences for Fair Labour Certification," MPRA Paper No. 73718 (2016), https://core.ac.uk/download/pdf/213987413.pdf.

56. This fundamental challenge is also a core reason there is so much more regulation today than in previous eras. Anderson, "Liberty, Equality, and Private Government," March 4–5, 2015, https://tannerlectures.utah.edu/_resources/documents/a-to-z/a/Anderson%20manuscript.pdf.

57. Bernard Kilian et al., "Can the Private Sector Be Competitive and Contribute to Development through Sustainable Agricultural Business? A Case Study of Coffee in Latin America," *International Food and Agribusiness Management Review* 7, no. 3 (2004).

58. See, for example, Abby Nájera and Homero Fuentes, *Child Labor, Forced Labor and Land Rights & Use: Belize Sugar Cane Supply Chain Country Study for The Coca-Cola Company Report Harvest 2016–2017*, Commission for the Verification of Corporate Codes of Conduct (April 12, 2018).

59. Christopher Cramer et al., *Fairtrade, Employment and Poverty Reduction in Ethiopia and Uganda*, UK Department for International Development (April 2014), https://www.soas.ac.uk/ftepr/publications/file144219.pdf.

60. Ibid.

61. See, for example, David Levy, Juliane Reinecke, and Stephan Manning, "The Political Dynamics of Sustainable Coffee: Contested Value Regimes and the Transformation of Sustainability," *Journal of Management Studies* 53, no. 3 (April 17, 2016): 364–401.

62. Pippa Stevens, "ESG Index Funds Hit $250 Billion as Pandemic Accelerates Impact Investing Boom," CNBC, September 2, 2020, https://www.cnbc.com/2020/09/02/esg-index-funds-hit-250-billion-as-us-investor-role-in-boom-grows.html.

63. NASA, "The Effects of Climate Change," accessed June 25, 2021, https://climate.nasa.gov/effects/.

64. Torsten Ehlers, Benoit Mojon, and Frank Packer, "Green Bonds and Carbon Emissions: Exploring the Case for a Rating System at the Firm Level," *BIS Quarterly* (September 14, 2020), https://www.bis.org/publ/qtrpdf/r_qt2009c.htm.

65. Ibid.

66. Ibid.

67. U.S. Government Accountability Office, Report to the Honorable Mark Warner of the U.S. Senate, GAO-20-530, "Public Companies: Disclosure of Environmental, Social, and Governance Factors and Options to Enhance Them" (July 2020), https://www.gao.gov/assets/710/707949.pdf.

68. Luluk Widyawati, "A Systematic Literature Review of Socially Responsible Investment and Environmental Social Governance Metrics," *Business Strategy and the Environment* 29, no. 2 (February 2020): 619–37; Sakis Kotsantonis and George Serafeim, "Four Things No One Will Tell You About ESG Data," *Journal of Applied Corporate Finance* 31, no. 2 (Spring 2019): 50–58.

CHAPTER 9: CONNECTIONS, LOCAL AND GLOBAL

1. Lewis Hyde, *The Gift: How the Creative Spirit Transforms the World*, 3rd ed. (New York: Vintage Books, 2019).

2. Ibid., 72.

3. James Laube, "Not Just Another Anderson Valley Roadside Attraction," *Wine Spectator*, May 13, 2013, https://www.winespectator.com/articles/not-just-another-anderson-valley-roadside-attraction-48422.

4. "What Is a Bottle of Wine," Covenant Blog, accessed October 29, 2021, https://covenantwines.com/wine/truth-in-wine-what-is-a-bottle-of-wine-worth/; Tom Wark, "This Is Why Alcohol Self-Distribution Is the Next Big Thing," *Fermentation*, March 2, 2020, https://fermentationwineblog.com/2020/03/this-is-why-alcohol-self-distribution-is-the-next-big-thing/.

5. "A Great Case of Navarro Wine," Dan Dawson's Wine Advisor, accessed March 1, 2021, https://dawsonwineadvisor.com/great-case-of-navarro-wine.

6. Dorothy J. Gaiter and John Brecher. "Infinite Tastes in a Unicorn Wine: The Story of Navarro's Dry Gewürztraminer," *Grape Collective*, October 11, 2019, https://grapecollective.com/articles/infinite-tastes-in-a-unicorn-wine-the-story-of-navarros-dry-gewrztraminer.

7. Interview with the author, December 16, 2020.

8. "2018 Pinot Noir," Navarro Vineyards, https://www.navarrowine.com/shop/2018-pinot-noir-methode-a-l'ancienne.

9. Lawrence Lessig, *Code: And Other Laws of Cyberspace* (New York: Basic Books, 1999).

10. Interview with the author, December 9, 2020.

11. Steven Brown, "The COVID-19 Crisis Continues to Have Uneven Economic Impact by Race and Ethnicity," Urban Institute, July 1, 2020, https://www.urban.org/urban-wire/covid-19-crisis-continues-have-uneven-economic-impact-race-and-ethnicity.

12. Courtney Higgs, "Indie Beauty Brands Need Us More Than Ever—Here Are 36 You Can Support Now," Who What Wear, May 13, 2020, https://www.whowhatwear.com/indie-beauty-brands-to-support.

13. Dana Mattioli, "Not Being Amazon Is a Selling Point for These E-Commerce Players," *Wall Street Journal*, June 16, 2021, https://www

.wsj.com/articles/not-being-amazon-is-a-selling-point-for-these-companies
-11623835802.

14. Margaret Atwood, "The Gift of Lewis Hyde's 'The Gift,'" *Paris Review*, September 16, 2019.

15. Hyde, *Gift*, 27.

16. Ibid., 78.

17. Elena Renken, "Most Americans Are Lonely, and Our Workplace Culture May Not Be Helping," NPR, January 23, 2020, https://www.npr.org/sections/health -shots/2020/01/23/798676465/most-americans-are-lonely-and-our-workplace -culture-may-not-be-helping; Cigna Newsroom, "Cigna Takes Action to Combat the Rise of Loneliness and Improve Mental Wellness in America," Cigna, January 23, 2020, https://newsroom.cigna.com/cigna-takes-action-to -combat-the-rise-of-loneliness-and-improve-mental-wellness-in-america.

18. Jena McGregor, "This Former Surgeon General Says There's a 'Loneliness Epidemic' and Work Is Partly to Blame," *Washington Post*, October 4, 2017, https://www.washingtonpost.com/news/on-leadership/wp/2017/10/04 /this-former-surgeon-general-says-theres-a-loneliness-epidemic-and-work-is -partly-to-blame/?utm_term=.3711f93a6291.

19. Ibid.

20. Raghuram Rajan, *The Third Pillar: How Markets and the State Leave the Community Behind* (New York: Penguin Press, 2019), 10–11.

21. Interview with the author, December 10, 2020.

22. "Photo of Breonna Taylor," Hanahana Beauty, Instagram, September 24, 2020, https://www.instagram.com/p/CFiR8V8lZrc/.

CHAPTER 10: ALMOST-DIRECT, QUASI-DIRECT, AND THE LIMITS OF DIRECT

1. Some of this material was first published in Kathryn Judge, "The Future of Direct Finance: The Diverging Paths of Peer-to-Peer Lending and Kickstarter," *Wake Forest Law Review* 50 (2015): 603–42.

2. Annys Shin, "Want to Loan Me Money? Here's a Picture of My Dog," *Washington Post*, January 27, 2007, https://www.washingtonpost.com/archive /business/2007/01/27/want-to-loan-me-money-heres-a-picture-of-my-dog -span-classbankheadprosper-links-people-who-need-money-with-those-who -have-it-to-lendspan/69f0ba0a-2045-4bd4-b49f-0ff2d4247181/.

3. "Banks Watching Latest Online Trend: Strangers Asking Strangers for Loans," *Times Trenton*, November 28, 2007, at A15.

4. Paul Katzeff, "Pros, Cons of Peer-to-Peer Lending," *Investor's Business Daily*, June 12, 2009, 2:15 p.m., http://news.investors.com/investing-mutual -funds/061209–479444-pros-cons-of-peer-to-peer-lending.htm.

5. Devin G. Pope and Justin R. Sydor, "What's in a Picture? Evidence of Discrimination from Prosper.com," *Journal of Human Resources*, no. 46 (Winter 2011): 53–92.

6. Taylor Moore, "Peer-to-Peer (P2P) Lending," NextAdvisor, December 8, 2020, https://time.com/nextadvisor/loans/personal-loans/peer-to-peer-lending/.

7. Ibid.

8. Nav Athwal, "The Disappearance of Peer-to-Peer Lending," *Forbes*, October 14, 2014, http://www.forbes.com/sites/groupthink/2014/10/14/the-disappearance -of-peer-to-peer-lending/.

9. Ruby Hinchliffe, "LendingClub Shuts Retail P2P Offering as It Focuses on Institutional Investors," FintechFutures, October 9, 2020, https://www .fintechfutures.com/2020/10/lendingclub-shuts-retail-p2p-offering-as-it -focuses-on-institutional-investors/; Hannah Lang, "OCC Approves Lending-Club Acquisition of Radius," *American Banker*, December 31, 2020, https:// www.americanbanker.com/news/occ-approves-lendingclub-acquisition-of -radius.

10. Leonard Schlesinger, Matt Higgins, and Shaye Roseman, "Reinventing the Direct-to-Consumer Business Model," *Harvard Business Review Digital Articles*, March 31, 2020.

11. Lawrence Ingrassia, *Billion Dollar Brand Club: How Dollar Shave Club, Warby Parker, and Other Disruptors Are Remaking What We Buy* (New York: Henry Holt and Company, 2020).

12. *Everlane*, accessed March 23, 2021, https://www.everlane.com/factories /cashmere.

13. "Everlane," *FastCompany*, accessed October 29, 2021, https://www.fastcompany .com/company/everlane; Jessica Testa, Vanessa Friedman, and Elizabeth Paton, "Everlane's Promise of 'Radical Transparency' Unravels," *New York Times*, July 26, 2020, https://www.nytimes.com/2020/07/26/fashion/everlane-employees -ethical-clothing.html.

14. Zoe Schiffer, "Everlane Customer Experience Workers Say They Were Illegally Laid Off," *Verge*, April 2, 2020, https://www.theverge.com/2020/4/2/21069279/ everlane-customer-experience-union-majority-illegal; Whitney Bauck, "Former Everlane Employees Call Out Alleged Racism and Toxic Culture in the Workplace," Fashionista, June 24, 2020, https://fashionista.com/2020/06 /everlane-racism-toxic-workplace-culture?li_source=LI&li_medium=m2m -rcw-fashionista; Whitney Bauck, "Former Everlane Employees Claim They Were Unlawfully Fired After They Tried to Unionize [Updated]," Fashionista, August 17, 2020, https://fashionista.com/2020/04/everlane-union-bust -covid-19.

15. Testa, Friedman, and Paton, "Everlane's Promise."

16. Ingrassia, *Billion Dollar Brand Club*, 291.

17. Karen Iorio Adelson, "Are the Bras from These 7 New(ish) Start-ups Actually Good?" *Strategist*, November 14, 2019, https://nymag.com/strategist/article /start-up-bra-reviews-thirdlove-lively-true-and-co.html.

18. Zoe Schiffer, "ThirdLove Says It's by Women, for Women. But Women Who've Worked There Disagree," *Vox*, September 16, 2019, https://www.vox.com /the-goods/2019/9/16/20864206/thirdlove-bra-company-women-employees -quit-ceo; Zoe Schiffer, "Emotional Baggage," *Verge*, December 5, 2019, https:// www.theverge.com/2019/12/5/20995453/away-luggage-ceo-steph-korey -toxic-work-environment-travel-inclusion.

19. Ethan Wolff-Mann, "Here's How to Buy Dollar Shave Club Razors for Less Money on Amazon," Money.com, July 20, 2016, https://money.com/dollar -shave-club-razors-dorco-amazon/.

20. Tom Foster, "Over 400 Startups Are Trying to Become the Next Warby Parker. Inside the Wild Race to Overthrow Every Consumer Category," *Inc.*, May 2018, https://www.inc.com/magazine/201805/tom-foster/direct-consumer -brands-middleman-warby-parker.html.

21. The rise and dominance of today's digital platforms raise a host of issues beyond those addressed here. For a helpful primer of some of the core issues and their implications, see Rory Van Loo, "Federal Rules of Platform Procedure," *University of Chicago Law Review* 88 no. 4 (2021): 829-96.

22. Jean-Charles Rochet, "Two-Sided Markets: A Progress Report," November 29, 2005, http://publications.ut-capitole.fr/1207/1/2sided_markets.pdf (provides a good primer on these markets and where they fit into the economics literature).

23. Aron Hsiao. "How to Calculate Ebay and PayPal Fees," *The Balance Small Business*, June 17, 2020, https://www.thebalancesmb.com/what-to-know -about-ebay-and-paypal-fees-before-you-sell-1140371.

24. Grace Dobush, "How Etsy Alienated Its Crafters and Lost Its Soul," *Wired*, February 19, 2015.

25. Julia Brucculieri, "Here's What It's Really Like to Make a Living Selling on Etsy," *Huffington Post*, November 22, 2018, https://www.huffpost.com/entry /how-to-make-money-etsy-secrets_n_5be9f95ee4b0caeec2bc9e91.

26. Chad Dickerson, "Notes from Chad," *Etsy News*, November 18, 2011, https://blog.etsy.com/news/2011/notes-from-chad-4/.

27. Brucculieri, "Selling on Etsy."

28. Ilyssa Meyer, "Etsy Celebrates the Creative Entrepreneurs We Support Across the Globe," *Etsy News*, April 29, 2019, https://blog.etsy.com/news/2019/etsy -celebrates-the-creative-entrepreneurs-we-support-across-the-globe/.

29. Thomas Philippon, *The Great Reversal: How America Gave Up on Free Markets* (Cambridge, MA: Harvard University Press, 2019).

30. Dennis R. Shaughnessy, "The Public Capital Markets and Etsy and Warby Parker," *SEI at Northeastern*, October 10, 2018, https://www.northeastern .edu/sei/2018/10/the-public-capital-markets-and-etsy-and-warby-parker/.

31. Interview with the author, July 20, 2021.

32. Ethan Mollick, "Containing Multitudes: The Many Impacts of Kickstarter Funding," 2016, https://papers.ssrn.com/sol3/papers.cfm?abstract_id=2808000 &download=yes.

33. Wayne Duggan, "This Day in Market History: The Dell IPO," Benzinga, June 22, 2021, https://www.benzinga.com/general/education/21/06/11921134/this -day-in-market-history-the-dell-ipo.

CHAPTER 11: FIVE PRINCIPLES FOR POLICY MAKERS, COMPANIES, AND THE REST OF US

1. Jules Struck, "Crisp Air and Apples: Pandemic-Weary Folks Flock to Pick-Your-Own Farms," *Christian Science Monitor*, October 8, 2020, https:// www.csmonitor.com/The-Culture/Food/2020/1008/Crisp-air-and-apples -Pandemic-weary-folks-flock-to-pick-your-own-farms; Nancy Shohet West, "Apple Picking Time Seems Sweeter Than Ever," *Boston Globe*, September 17,

2020, https://www.bostonglobe.com/2020/09/17/metro/apples-are-safe-pick-during-pandemic/.

2. Lawrence A. Cunningham and Stephanie Cuba, *Margin of Trust: The Berkshire Business Model* (New York: Columbia University Press, 2020).

3. Securities and Exchange Commission, "Fund of Funds Arrangements," 85 Fed. Reg. 73,924 (codified at 17 C.F.R. pts. 270, 274) (November 19, 2020).

4. Barr, Jackson, and Tahyar, *Financial Regulation* and sources cited therein.

5. Ibid.

6. Mark A. Cohen, "Imperfect Competition in Auto Lending: Subjective Markup, Racial Disparity, and Class Action Litigation," *Review of Law & Economics* 8, no. 1 (2012): 21–58.

7. Kerwin Kofi Charles, Erik Hurst, and Melvin Stephens Jr., "Rates for Vehicle Loans: Race and Loan Source," *American Economic Review* 98, no. 2 (2008): 315–20, at 317.

8. Lisa Rice and Erich Schwartz Jr., *Discrimination When Buying a Car: How the Color of Your Skin Can Affect Your Car-Shopping Experience*, National Fair Housing Alliance Report (2018).

9. Alison DeNisco Rayome, "Best Food Delivery Service," CNET, May 13, 2020, https://www.cnet.com/news/best-food-delivery-service-doordash-grubhub-uber-eats-and-more-compared/.

10. *Granholm v. Heald*, 544 U.S. 460 (2005); *Costco Wholesale Corp. v. Hoen*, 2006 U.S. Dist. LEXIS 27141, 2006 WL 1075218 (W.D. Wash., April 21, 2006), affirmed in part by 538 F.3d 1128 (9th Cir. 2008); *Ass'n of Washington Spirits & Wine Distributors v. Washington State Liquor Control Bd.*, 182 Wash. 2d 342 (2015).

11. Martin J. Gruber, "Another Puzzle: The Growth in Actively Managed Mutual Funds," *Journal of Finance* 51, no. 3 (1996): 783–810; Eugene F. Fama and Kenneth R. French, "Luck versus Skill in the Cross-Section of Mutual Fund Returns," *Journal of Finance* 65, no. 5 (2010): 1915–47.

12. Patrick McGeehan, "Panel's Report Offers Details on 'Spinning' of New Stocks," *New York Times*, October 3, 2002, https://www.nytimes.com/2002/10/03/business/panel-s-report-offers-details-on-spinning-of-new-stocks.html.

13. Tamar Adler, "Diaspora Co.'s Fair-Trade Spices Will Enlighten More Than Your Cooking," *Vogue*, April 9, 2021, https://www.vogue.com/article/diaspora-co-fair-trade-spices.

14. Khan, "Amazon's Antitrust Paradox."

15. Judge, "Fragmentation Nodes."

16. Judge, "The Future of Direct Finance."